Lighthouse Men & Women of the Moor

JAMES CARRON

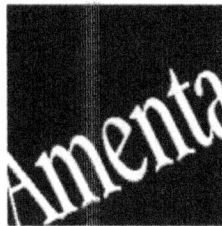

PUBLISHING

Amenta Publishing

amenta.ink

Lighthouse Men & Women of the Moor

By James Carron

First published 2021 by Amenta Publishing

Copyright © James Carron 2021

ISBN 9798525595262

Front cover: Corrour Station (Author) and snow plough at Corrour Station, 1894 (John Alsop)

Rear cover: First ScotRail Class 156 SuperSprinter 156476 heads north out of Corrour towards Corrour Summit, 2010 (Author)

CONTENTS

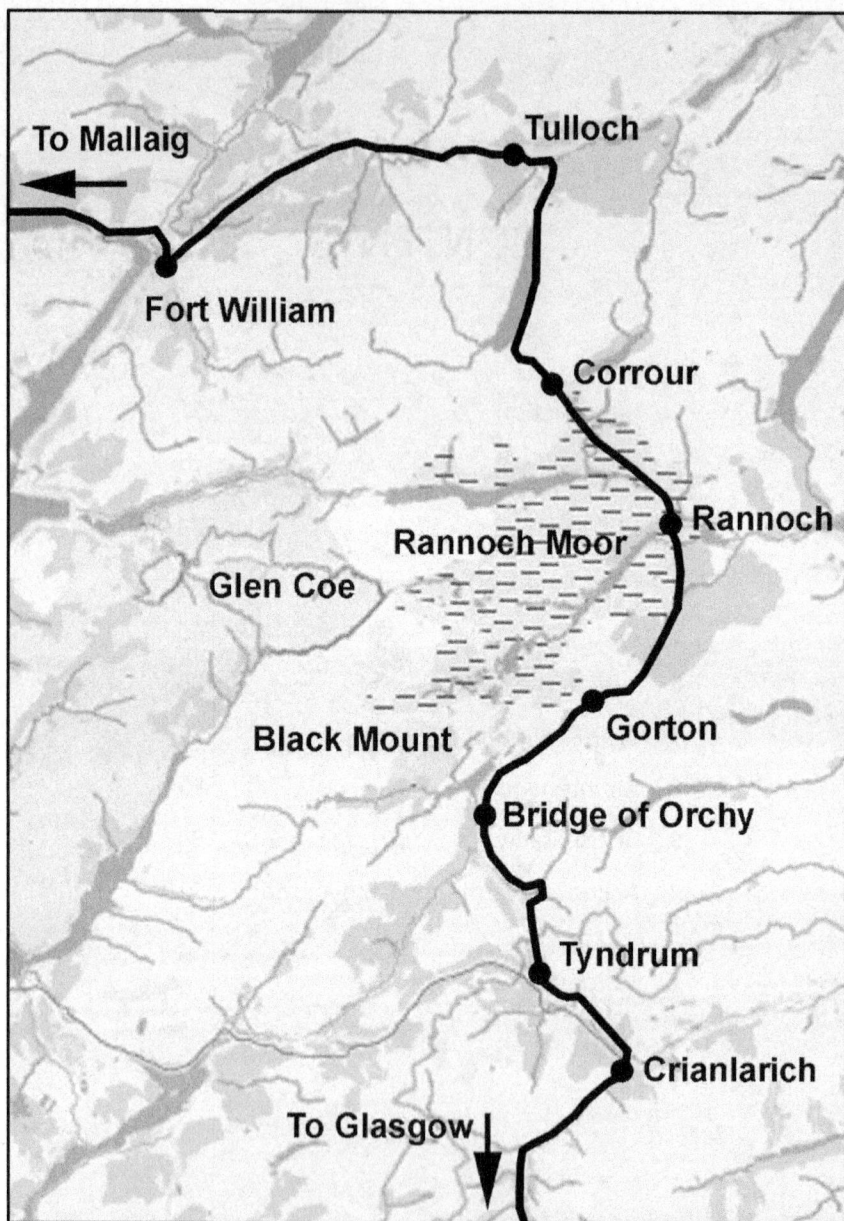

A map showing the route of the northern portion of the West Highland Railway, crossing Rannoch Moor on its journey from Crianlarich to Fort William

1

Introduction

The signalmen and women and stationmasters of Rannoch Moor worked in such far-flung, isolated outposts that 'lighthouse men and women of the moor' seems a particularly apt title for them. They looked after Corrour Siding, Rannoch Station and Gorton Crossing, all three remote beacons for the early Victorian and Edwardian travellers who ventured north to explore this barren wilderness.

It is not entirely original but rather a corrupted version of a quote I first read in the seminal history of the West Highland Railway, written by John Thomas and first published in 1965. He described them as 'lighthouse men of the land', a line that has resurfaced from time to time in magazine articles, newspaper reports and in a television documentary aired in 1987, a time when the railway and its employees were undergoing a significant period of change.

In keeping with the lighthouse men of the time, the staff of Corrour, Rannoch and Gorton lived lonely lives, separated from society, but their roles were vital to the safe and efficient running of the railway. Conditions were harsh and primitive, particularly in the early years. There was no electricity until the 1980s and, in the case of Gorton, no running water. Homes were heated and meals cooked with coal, paraffin lanterns and later batteries proffered lighting and they relied on the railway for supplies, which in times of bad weather could be severely delayed, leaving cupboards empty and bellies rumbling.

But, when the time came to leave, the last remaining railwaymen and women refused to go, such was their loyalty to the line and to the moor. It was a way of life that intrigued and often mystified passengers as they traversed this bleak back country. Rather than simply offer up another history of the West Highland Railway, this book aims to tell their story and where better a place to start than on the great moor itself.

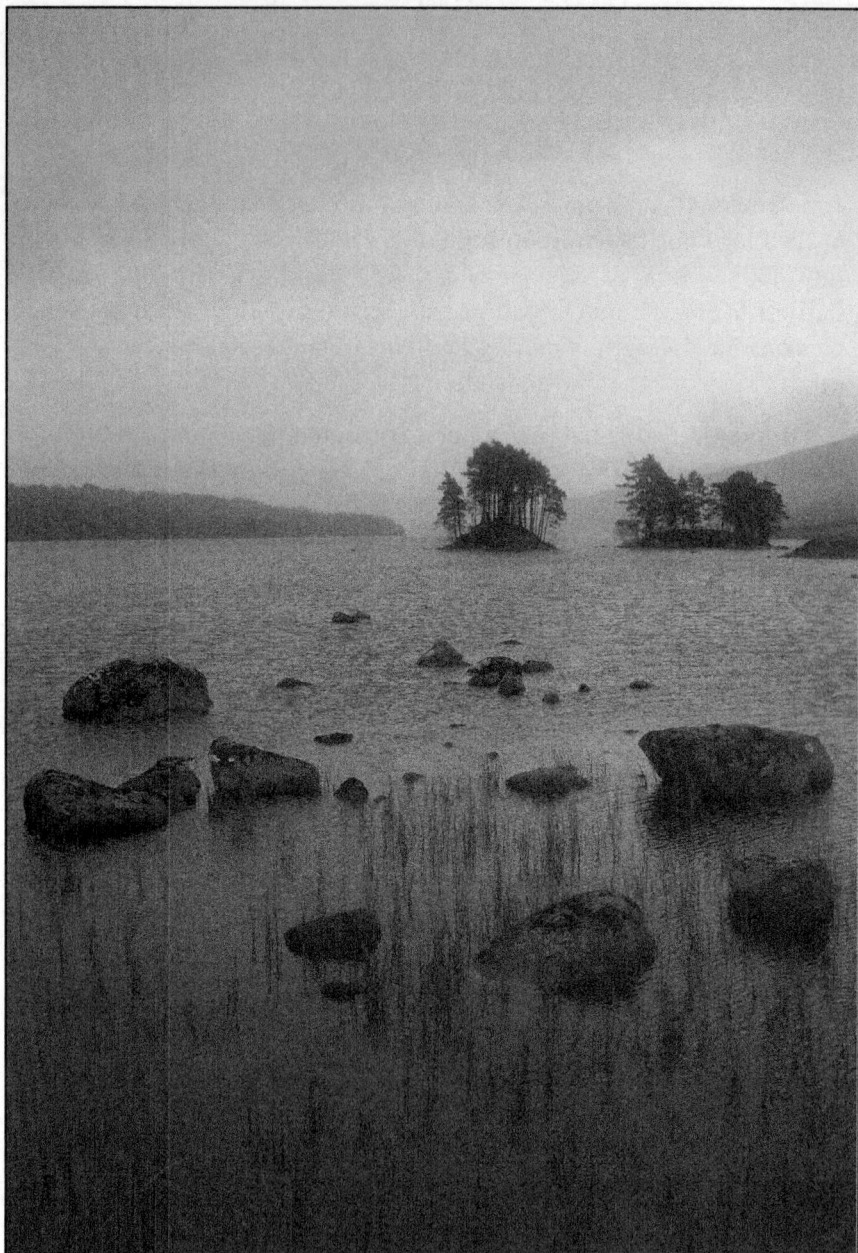

Loch Ossian, near Corrour Station (Author)

2

Rannoch Moor

Rannoch Moor is remarkably unpredictable. It is barren and blessed with a character that changes with the weather. On a bright sunny day, it is a land of striking natural beauty. On a bad day, it is dark and forbidding. In his novel, *Kidnapped*, Robert Louis Stevenson wrote: 'A wearier looking desert a man never saw'. Desert is perhaps an ironic choice of words for, thanks to its unique geography and geology, the moor is one of the wettest places in Britain.

Traversing this 50 square mile chunk of wilderness is no easy task and one that has challenged not only those on foot, bicycle or pony, but also engineers. It is a terrifyingly vast and featureless environment where the human form loses all significance. The landscape reigns supreme and nature has successfully warded off all but a handful of attempts to conquer it.

The only main arteries over the moor are the West Highland Railway, which opened in 1894, and the A82 road, which dates from the 1930s. Crossing on foot requires careful route planning, excellent navigational skills, and preparation for all weathers, as a group of Victorian railway surveyors found to their cost.

In January 1889, prior to the construction of the line, seven men set out to trace the railway's route across Rannoch Moor. The story of their adventure was recounted in an article published in *Blackwell's Magazine* in 1927. Entitled *Benighted on the Moor of Rannoch*, it was penned by assistant engineer Mr J E Harrison.

Departing from Spean Bridge, they took a boat down Loch Treig and at dawn the following day trudged track and bare moor to the site of Corrour Station. Harrison wrote: 'The moor extended in front in an apparently limitless expanse of peat moss and hummocky ridges of scattered boulders. A more desolate scene it would be difficult to picture.'

His description remains true to this day, as anyone caught out in bad weather will testify. Battling through wind and rain, the surveyors made it to the site of Rannoch Station but, cold and exhausted, they ran into difficulties and ended up stuck on the moor in freezing conditions. But more of their story later.

A north-south traverse is exceptionally difficult, due to the boggy terrain and profusion of lochs, rivers and streams. In freezing winter conditions, it is possible to push on to Gorton across the area of the black lochs but for the rest of the year this is an impenetrable quagmire.

Keen to experience the majesty of the moor in the depths of winter for myself and sample just a little of what life was like for the navvies who laboured on the railway and for the families who worked its stations and signal boxes, I embarked upon my own expedition, skirting the fringes on a three-day trek that would take me from Bridge of Orchy, in the south, to Corrour, the highest mainline railway station in Britain.

The moor is framed by a series of mountain ranges that funnel it west towards Glencoe, east to Loch Rannoch and south to the Black Mount. During the Ice Age, it was covered by an extensive ice cap, glaciers radiating out across the West Highlands. After the ice melted, water remained, the gentle slope, slow drainage and high rainfall carpeting the underlying granite with blanket bog. Until Roman times it was partially wooded but today it is virtually treeless, save for small clumps on loch islands and stray silver birches clinging to the line of the railway.

Rannoch Moor in winter (Author)

King's House Hotel (Author)

This lack of natural shelter, combined with a dearth of settlements, does not encourage visitors to stray far from the beaten track. To do so would invite only isolation and exposure. Thankfully, for the first leg of my route, a section of the West Highland Way, Scotland's most popular long-distance hiking trail, offered sound footing, leading me from Bridge of Orchy across the Black Mount to King's House.

Beyond Forest Lodge, near Inveroran, the way follows Thomas Telford's 19th century road. As I ventured out on to the expansive wastelands of the Black Mount, it was breathtakingly bleak, horizon and skyline converging in a grey ribbon of despair. I plodded on through the tumbling rain and the numbing cold. The sky, the ground, my feet, they were all wet. Slowly I was nearing saturation point. My body felt like an antiquated central heating boiler, radiating what little warmth it could muster in a fruitless bid to heat the moisture circulating through my boots and clothes. I had expected Rannoch Moor to be wet, but the elements exceeded all expectations.

Ba Bridge marked the midpoint of my Black Mount crossing. Moor lochs to the east were barely visible through the cloud and the River Ba flowed ferociously beneath my feet, its torrents swelled by the downpour. From here, I squelched north towards Glencoe, crossing what Harrison described as a 'desolate isle in a waste of vapour'.

Plans to penetrate the West Highlands with railways were first mooted in the 1840s, a time when railway mania was at its peak in Scotland. The more grandiose schemes, including one to span Loch Ness with a mighty

5500-foot-long viaduct and tunnel through Ben Nevis, attracted excited attention but few backers and came to nothing.

In was 1880 before more realistic proposals were drawn up for a line – the Glasgow and North Western Railway – that would run north from the second city of the Empire by Loch Lomond to Crianlarich where the route would turn west, crossing the Black Mount beyond Bridge of Orchy and pausing at King's House before descending through Glen Coe to Ballachulish. From here, it would run by Loch Linnhe to Fort William before passing through the Great Glen to Inverness.

Arriving at the King's House Hotel, I was drenched. Thankfully, a tumble dryer dealt with my wet clothes and a good meal and a few drinks at the bar revived my soggy spirits before I turned in for the night.

Established in the 17th century, the King's House is one of Scotland's oldest licensed inns. In 1746, government troops tasked with hunting down fleeing Jacobites were billeted here following the Battle of Culloden and in subsequent years it served travellers crossing from Rannoch Moor into Glen Coe. It was also a popular haunt for navvies employed on the West Highland Railway and later the Blackwater Dam, near Kinlochleven.

If the experiences of Dorothy Wordsworth – sister of poet William Wordsworth – are anything to go by, it was a far from salubrious joint. Visiting in 1803 she wrote: 'Never did I see such a miserable, such wretched place, long rooms with ranges of beds, no other furniture except benches, or perhaps one or two crazy chairs, the floors far dirtier than an ordinary house could be if it were never washed.'

Long popular with walkers and climbers, the King's House was always homely and welcoming, if a little rough around the edges. It reminded me of visits to the home of an elderly aunt who occupied a rambling Victorian terrace house, where paint flaked and floorboards creaked.

For those who arrived sodden after a wet and cold day in the great outdoors, the drying room and temperamental tumble drier were heaven sent while shared bathrooms offered showers, baths and a steady flow of hot water.

Packed with character and curiosity but long overdue a revamp, the hotel was eventually taken in hand and, while retaining its ancient heart, the 1960s extension was demolished and replaced with a contemporary new wing in 2018. The bunkhouse was also rebuilt. Today, it is a very modern, well-equipped hotel, offering levels of comfort previously unheard of in this remote spot.

The following morning, there was a glimmer of sun in the sky and the world seemed a more welcoming place. Dry, warm and well fed, I was out early, tramping the track to Black Corries Lodge with a spring in my step. The previous day's watery misery was but a distant memory and I made good progress, following the lively River Etive upstream.

At the lodge, the vast scale of Rannoch Moor came sharply into focus. From my elevated viewpoint, I surveyed a vast plain of brown grass and heather, liberally sprinkled with lichen and moss-encrusted boulders. Small lochans dotted a landscape where rivers and streams run riot.

Due to the sheer volume of water emptying itself off the hillside, underfoot conditions rapidly deteriorated as I walked east from the lodge. While initial attempts to hop from one tussock of grass to another proved relatively successful, there was ultimately no escaping the slavering jaws of the moor and I soon found myself ankle deep in mud and water.

Progress slowed to little more than a crawl in places but after a long battle with the bog, and guided by a line of telegraph poles, I finally entered woodland above Loch Laidon.

After spending so much time on open ground, the hillside plantation felt claustrophobic and rather at odds with the moorland ethos of my hike so there was no reluctance to leave the trees when the track finally emerged from their confines at the northern end of the loch. The terrain improved dramatically, pushing me on to Rannoch Station and a night at the Moor of Rannoch Hotel.

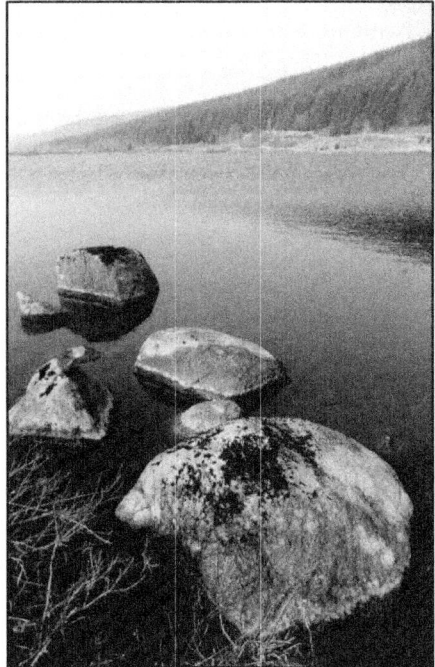

Loch Laidon (Author)

In his book *Iron Road to the Isles*, author Michael Pearson writes of Rannoch Station: 'There's no signal for your mobile phone, you cannot be got at, an invigorating feeling of optimism and potential manifests itself in your mind. However long you are here for it is not time to be killed, but time to be treasured.'

Moor of Rannoch Hotel (Author)

The tiny hamlet (population: circa 5) owes its existence to the railway. Aside from the station and hotel, there is a telephone box and a couple of cottages. Now unmanned, the station houses a cosy little tearoom and the Rannoch Moor Visitor Centre. Display boards outline the history, geology and wildlife of the moor and there is a fascinating section on the construction of the railway. So spongy was the terrain that the inventive Victorian engineers were forced to lay down a mat of wood, effectively floating the line across the bog.

The party of railway surveyors followed the course of the line from Corrour, to the north, to here. However, I planned to join the firmer Road to the Isles for the final leg of my walk to Corrour.

Rannoch Station, showing the signal box and station building (Author)

I set off east along the narrow road linking Rannoch Station with civilisation and continued as far as Loch Eigheach where the old drove road strikes north, skirting below Meall na Mucarach before rising over the western flank of Carn Dearg. Close to its highest point, the route passes the stark ruins of Corrour Old Lodge, once the highest shooting lodge in Scotland. This is a fine vantage point from which to view the moor.

The ruins of Corrour Old Lodge (Author)

It really is a terrifyingly vast and featureless place, an apparently limitless expanse of peat moss and grassy ridges scattered with huge boulders. Some describe it as desolate, others as a wasteland. But hit it on a good day and it is an enchanting and mesmerising place. You can immerse yourself in the landscape, walk freely without meeting another soul. It refreshes and revives body and soul and there is a level of peace and tranquillity that would be hard to find anywhere else in Britain.

While the railway surveyors found only desolation at Corrour, present day visitors will find sanctuary – and a surprising level of comfort – in a restaurant and café, boutique bed and breakfast rooms and youth hostel.

The latter is a real retreat. Staying here, in this simple wooden ex-boathouse, is a delightful back to basics affair with none of the conveniences many take for granted. The spectacular location, however, more than compensates and the place has excellent green credentials.

Loch Ossian Youth Hostel (Author)

In 2003 it was refurbished as one of the UK's first eco-hostels. A wind turbine provides power and there are ecologically sound water and waste disposal systems.

Up at Corrour Station, the cosy restaurant provided a warm and welcome end to my Rannoch Moor adventure. As I waited for the next train south, I sat by a roaring log fire and savoured a hot bowl of soup.

Luxury was the last thing I had expected on my foray into this wild corner of Scotland. But, in every desert, there is an oasis, and, after three days on the moor, I had at last found mine. And it owed its existence to the railway.

The Victorian railway surveyors were less fortunate, finding no vestiges of refuge at Corrour. There was no roaring fire and no hot bowls of soup for them and they were forced to plod on. Unlike my adventure, which began in the south, their trek started to the north-west of Corrrour, in Fort William.

Running into significant opposition, proposals for the Glasgow and North Western Railway were rejected after two months of parliamentary debate and a new scheme for the West Highland Railway was drawn up. Backed by the North British Railway Company (NBR), which intended to operate the line, the project was less ambitious, at 99 miles in length, as opposed to its predecessor's tally of 167 miles.

The modern day incarnation of Corrour Station House (Author)

The new line would track the same course north from Glasgow via Loch Lomond to Crianlarich where, once again, it would turn west. Beyond Bridge of Orchy, however, rather than progress over the Black Mount, it would curve north, crossing Rannoch Moor, before approaching Fort William from the east, through Glen Spean. Thereafter, there was scope to add branches, and an extension to Inverness remained a distinct possibility.

With only five wealthy landowners to placate, progress was more positive and, with the project gathering steam, Robert McAlpine, one of Britain's greatest entrepreneurs of the time and a pioneer in railway construction, was keen for a slice of the action. He travelled north to throw his hat into the ring.

Keen to see the terrain for himself, McAlpine joined forces with six other

Robert McAlpine (National Portrait Gallery)

15

men for the hike over the moor, their intention to collect evidence on the lie of the land and its remoteness in support of the NBR's West Highland Railway Bill.

Three members of the party were from the Glasgow civil engineering firm of Formans & McCall – Charles Forman, who had been appointed lead engineer on the project, his chief engineer James Bulloch, and assistant engineer J E Harrison. They were joined by John Bett, factor for Breadalbane Estates, Major Martin of Kilmartin, factor for Poltallock Estates, and Mr N B MacKenzie, a lawyer from Fort William.

Charles Forman

in its original form, a veil of secrecy was cloaked over the participants. None was named in Harrison's account of events published in *Blackwell's Magazine*. It was only years later that their identities were revealed.

Assembling for departure, the party was not particularly well equipped for such an ambitious winter expedition. Harrison described Bett, who was 60 years of age, as 'a townsman in appearance, equipped for the event in a high sided felt hat and waterproof and carrying an umbrella.' McAlpine, who was 41 at the time, he noted was a 'stout, full-blooded and loquacious man', with a hat.

Spending the night of January 29 in the comfort and warmth of a hotel in Spean Bridge, the group departed after breakfast the following morning, driving east to Inverlair Lodge, the home of Colonel Sir George Gustavus Walker, who owned neighbouring Corrour Estate.

There they enjoyed lunch before walking the two and a half miles south by gravel track to Loch Treig. The weather was none too promising – the day was cold and cloudy with a threat of sleet.

Their journey should have been straightforward.

Loch Treig (Author)

Plans were made to meet a boatman at the northern end of the loch who would row them the six miles south over the water to Creaguaineach Lodge, where they would find bed and board for the night.

But things began to go array when they reached the boathouse where they found neither boat nor boatman. With dusk falling and a storm brewing, the men attempted to break into the shed, prompting the sudden appearance of an angry man demanding to know what they were up to.

This, it transpired, was the boatman and, retrieving his craft, they set off into blustery winds and heavy rain. It soon became apparent that the boat was heavily overloaded and leaking badly. Sticking close to the eastern shoreline, the men frantically bailed out the water with their boots and paddled for over five hours, finally arriving at Creaguaineach Lodge at midnight.

Word had not, as planned, preceded them and there was no hearty meal to revive body and soul or warm beds to retire contentedly to. Some food and blankets were rustled up by the lodgekeeper and the party spent what remained of the night grabbing what sleep they could, most kept awake listening to the hail and sleet rattling against the window panes. It did not bode well.

Creaguaineach Lodge (Author)

Conditions were no better by morning. If anything, the weather was worse. But the railwaymen were undeterred and set off at dawn, the lodgekeeper guiding them to the end of the track where bare moor beckoned.

Although remote, the area was not uninhabited. Aside from the lodge, there was a small farm at Lochtreighead, to the east, and a steading at Luibruairidh, midway between Loch Treig and Loch Ossian, all linked by the Road to the Isles, a well-used but unsurfaced drove route. The party's course, however, took them south-east, away from these isolated outposts and over rough uncultivated ground.

A re-enactment of the expedition appeared in the 1970s film *A Line for All Seasons*, directed by Eddie McConnell, stills from which appear above, showing four of the men crossing the moor, and opposite.

Harrison wrote: 'The wind blew a gale and the day was darkened by low flying clouds. Sheets of sleet chased one another in raking columns, blotting out the horizon, and rendering the visible landscape as if it were a desolate isle in a waste of vapour.

'The moor extended in front in an apparently limitless expanse of peat moss and hummocky ridges of scattered boulders. A more desolate scene it would be difficult to picture.'

Passing below Leum Uilleim, the party found nothing at the proposed site of Corrour Station and was forced to tramp on. 'The going,' wrote Harrison, 'becomes very heavy and most exhausting.'

Following the course of the planned railway, the group proceeded slowly, passing to the west of the old Corrour Lodge and crossing the Black Water and then the county boundary to reach the River Gaur, half a mile to the south of where Rannoch Station would be located.

Here they were to meet landowner Sir Robert Menzies, of Rannoch Lodge, but he instead sent his head keeper out. The man suggested the group accompany him back to the shelter of the lodge for the night. With damp pocket watches now showing 1.30pm, they had perhaps three hours of daylight left and another eight miles of walking over trackless moor. The group decided to press on, a decision they would soon come to regret.

'Gradually the pace became slower and more laboured,' Harrison recounted. 'In a state of straggling disunion, the little group moved forward slowly until the elderly land agent (Bett), now far behind, stopped and said he could go no further.'

With Bett exhausted and two others not far from it, it was decided that Bulloch, who had crossed the moor on foot before during earlier survey work, would make his way to a shepherd's cottage at Gorton, seven miles away, and raise the alarm. Forman and MacKenzie would follow while the rest of the party remained on the moor.

Bullish McAlpine had other ideas. Harrison wrote: 'The contractor declared he would carry on by himself, adamant he was going to escape the moor.'

With only the shelter of their umbrellas and a bed of soggy grass, the stricken men spent several long, freezing hours on the moor, Bett lapsing in and out of consciousness. Their 'miserable and weary vigil' continued until midnight when two shepherds and a dog from Gorton, alerted by Bulloch, stumbled upon them.

The shepherd's cottage at Gorton (F Leask)

After administering a reviving dose of whisky, they guided the men two miles to a wooden hut where they took refuge by a roaring peat stove before setting off for Gorton the following morning.

A subsequent search found Forman and MacKenzie huddled together behind a boulder while McAlpine made it to the Inveroran Inn where he was found occupying 'the best bed'.

'Whatever his bodily frame suffered, his vigour of language was unabated, and his wandering was reviewed in a flood of disjointed phraseology, selected mainly at random from an unpublished vocabulary,' Harrison wrote.

'The grey dawn had found him hatless and drenched, bruised, dishevelled and wild gazing down into the misty valley from far above.'

Harrison noted that McAlpine appeared to be more worried about the loss of his hat than anything else. Of the seven men, he was the only one to successfully traverse the moor unaided.

Miraculously everyone survived and, thanks to the shepherds, the seven men were reunited at the Inveroran Inn. The next day, a fierce snowstorm swept the area. Harrison was in no doubt that had the blizzard come 24 hours earlier and hit the party while they were still out on the moor the expedition would have ended in tragedy.

Information gathered on the trek was duly submitted to parliament after which Harrison's account was filed away until 1927 when it was published in *Blackwood's Magazine*, a popular Edinburgh-based periodical that mixed fiction, satire, reviews and criticism.

A decade later, London and North Eastern Railway Company (LNER) director Arthur Murray had a copy of the article bound for the company archive and Harrison provided the names which were pencilled into the margins. The document passed to the British Railways Board and now resides with the National Records of Scotland, in Edinburgh.

Although McAlpine did not secure the contract for the West Highland Railway to Fort William, his firm did build the extension to Mallaig and it was here, through his pioneering use of concrete that he established his reputation for innovation, created one of his most famous engineering achievements, Glenfinnan Viaduct, and earned the nickname 'Concrete Bob'.

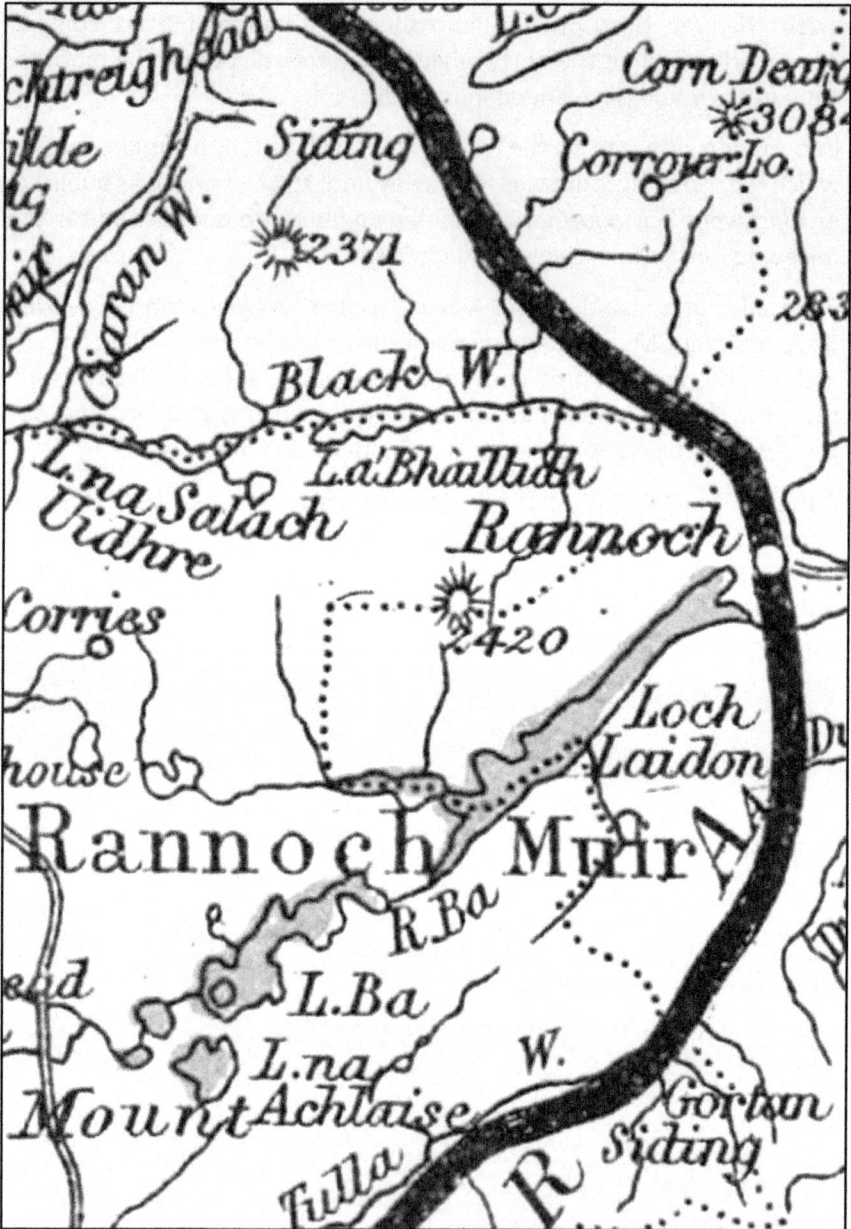

An extract from the Caledonian Railway Company's tourist map of 1910 showing rivals the North British Railway Company's West Highland Railway crossing Rannoch Moor

3
Making Tracks

The survival of Forman, Bulloch and Harrison, along with the passing of the West Highland Railway Act in July 1889, allowed construction of the line to commence and after the first ceremonial sod was cut in Fort William on October 23 of that year, contractors Lucas & Aird moved their materials and men in and set about realising Forman's plans for a 99-mile-long track linking Craigendoran Junction, at Helensburgh on the Firth of Clyde, with Fort William.

Construction began at five points and staging posts for the delivery of materials and camps for the workforce of predominantly Irish navvies were established.

The northern half, between Crianlarich and Fort William, was divided into two sections – Crianlarich to Gorton and Gorton to Fort William, the crossing of Rannoch Moor falling into the latter. In the north, navvies were based in Fort William, where materials were landed on a new pier built on Loch Linnhe, while, at the southern end, a camp was erected at Achallader, close to the northern end of Loch Tulla, on the road to Glen Coe. As progress was made, smaller intermediate camps popped up at Gorton and on the site of Rannoch Station.

The life of a navvy was a brutal one and it was no different for the labourers who toiled to drive the railway through the West Highlands with only basic hand tools and no heavy construction machinery.

Their only mechanical aid took the form of temporary standard gauge light railways laid to ferry the men and their tools and materials around.

Lucas & Aird was regarded as a good employer but deaths were common, most stemming from accidents at work (many involving the use of dynamite to blast rock); exposure and hypothermia (often linked to heavy alcohol consumption); fights (ditto); and the odd murder. There

Lucas & Aird

The company was a joint venture, formed in 1870 between brothers Charles and Thomas Lucas, who completed the Albert Hall, in London, in 1871, and John Aird & Company.

Both were London-based.

Charles De Neuville Forman

Born on August 10, 1852, Charles Forman was raised in Glasgow. He served his apprenticeship under his father, James Richardson Forman, who co-founded Forman & McCall, a company heavily involved in the development of the country's railways.

In 1876 the young engineer became a partner in the firm (renamed Formans & McCall) and he worked on various notable projects including the Kelvin Valley Railway, Glasgow Central Railway, Lanarkshire & Dunbartonshire Railway, Lanarkshire & Ayrshire Railway, Invergarry &Fort Augustus Railway and, of course, the West Highland Railway.

The heavy workload took its toll and Forman suffered a seizure and an attack of paralysis while in Spain in 1900. Thereafter his health steadily declined, and he died while resting and recuperating at Davos Platz, Switzerland, on February 8, 1901, at the age of just 48.

are no figures for the number of men who died building the West Highland Railway but in the area of Arrochar and Ardlui alone 37 deaths were recorded, the victims buried in and around a local cemetery where a memorial stands to mark both their contribution to the establishment of better transport links in the West Highlands, and their passing.

A report published in August 1891 revealed that there had been 26 deaths at nearby Tarbet during the previous winter, most the result of heavy drinking. While camps offering relative levels of comfort and security were offered by the contractors, some navvies, keen to save money, opted to provide their own accommodation, erecting simple wooden huts or tarpaulin tents, often without any form of heating. Falling asleep after a night on the bevvy, there was a real risk of succumbing to the cold, which many did. Others simply collapsed by the sides of roads, stumbling into ditches as they wandered home from the pub. Out for the count, hypothermia consumed them as they slept and, more often than not, their frozen bodies were found the following day by a passerby.

Over Christmas 1892 and into New Year 1893 there was an outbreak of smallpox at Ardlui. Investigations revealed that a tramp who stayed briefly in North Hut, one of the labourers' lodgings, had infected a number of navvies with the disease. Lucas & Aird segregated the men, fumigated the hut and burned all the bedding. The building was then turned into a makeshift isolation hospital, staffed by a nurse brought north from Glasgow. Of the 12 men who contracted the disease, two died.

A second outbreak, linked to the same tramp, was recorded in a navvies' hut down the line at Ardencaple, near Helensburgh, while a young girl who lost a leg when she fell under the wheels of a wagon on the line at Arrochar on December 28 was doubly unfortunate, falling ill with the disease. She survived after being treated in hospital in Glasgow. Another local girl, whose father had buried one of the smallpox victims, also succumbed to smallpox.

Between 2000 and, at the peak of construction, 5000 men were barracked in wooden huts in the West Highland Railway work camps. Each was equipped with medical facilities staffed by nurses who tended to both illness and injury.

Conditions and the weather ensured it was not a popular posting – the railway was referred to as the 'Long Drag' by navvies – and many spent only a short time in the area before moving on to find other work, which, at the time, was plentiful.

In 1908 Francis John Taylor penned an account of the Church of Scotland's outreach to the 10,000 or so navvies working on contracts around the country at that time which offered a fascinating insight into camp life.

He wrote: 'Navvies are among the strongest men in the country. Not an uncommon day's work shows the lifting of 15 to 20 tons of puddle by spadefuls or the wheeling of 60 tons of material in barrows a length of 20 yards.

'The old ideal of navvy life was summed up by one of their body as "hard work all the week for a full drink on Saturday afternoon and Sunday."

'To this might be added the further delights of dog washing, and hair cutting in the afternoon and fighting in the evening, as the alternative occupations of the navvy's day of rest.'

Taylor's missionary work took him to the work camps and huts of Scotland where he viewed at first hand the living conditions of the workmen and their families.

'The whole tendency of hut life may be put down as bad,' he observed. 'The huts are dirty, stuffy, and dusty to a very dangerous extent; let the hut-keeper be as cleanly as she will, she can scarcely keep them clean.

'In summer-time, when works are being pushed forward, the same beds are often occupied both night and day. The condition of the atmosphere, with the perfume of cooking, the breath of forty men, twenty of them in a drunken stupor, and not a window open, can be better imagined than described. It is surprising that the men are so healthy as they are.

'Yet they are brave, independent, generous, and noble in many of their unwritten laws; for while they would almost kill a policeman who ventured down the line to arrest a mate, they would give their last copper to a comrade in distress; and very few real navvies have ever been buried as paupers, and seldom do orphan children find a home in the poor-house.'

A group of navvies

The church established hostels and missions along the route of the new line to tend to the spiritual needs and welfare issues of the army of navvies while ministers opened libraries for the labourers in their manses, hoping to offer a distraction from drink.

Wages on the West Highland Railway ranged from 15 shillings a week for those employed south of Ardlui to 21 shillings a week for those working to the north. While pubs and stores selling alcohol claimed a good chunk of that, some of the pay was made in the form of food tokens to ensure it did not all go on drink.

Lucas & Aird employed a fleet of 32 steam locomotives in the project, 27 of their own and five borrowed from the NBR. To coordinate activities along the line and marshal the trains, between 30 and 40 telephones were installed at intervals along the route.

One of Lucas & Aird's steam locomotives in action during construction of the West Highland Railway

With fewer access points and greater engineering challenges, progress on the Crianlarich to Fort William section of the railway lagged behind the Craigendoran Junction to Crianlarich portion but by the end of June 1891 the line projecting east from Fort William had reached Tulloch, at the northern end of Loch Treig, and the railway board held a celebratory dinner to mark the achievement. It was a lavish affair but just a few weeks later all construction work halted, men were paid off and rumours began circulating that the line had been abandoned.

The crisis stemmed from a dispute between the board of the West Highland Railway Company and Lucas & Aird over money. As they progressed, the contractors discovered construction was more costly than they had initially estimated.

They attempted to renegotiate their contract and when the board refused to meet their demands, the case went to court. With the sheriff finding in favour of the railway company, it was back to the negotiating table and by October a new deal had been thrashed out, allowing work to begin again.

It was the spring of 1892 before sufficient men could be tempted back to enable real progress to be made and, on May 14 of that year, the *Glasgow Herald* reported on the state of play.

'The works were at once pushed on from Fort William and within a year about 10 miles of permanent way were laid as far as Spean Bridge,' the newspaper reported.

'Simultaneously, the formation of the track was proceeded with at other parts of the route, and several of the sections are now so nearly completed that they will be soon joined up.

'The general inaccessibility of the district through which the railway passes and the consequent difficulty of transporting material and of getting workmen at times have caused delays which have been increased by the rigours of the winter that has just passed. Nevertheless, all over good progress has been made and the contractors' trains are running on about one half, or 50 miles of the railway.'

Of the various breaks in the track that remained, the most significant was on Rannoch Moor, the crossing of which Charles Forman had first-hand experience of.

'From Crianlarich to some distance beyond Tyndrum contractors' trains are running,' the Glasgow *Herald* continued. 'There is here, however, a slight gap but beyond this the railway is again in operation on to the Achallader viaduct, opposite the Blackmount deer forest.

'On the Moor of Rannoch comparatively little work has been done and, roughly speaking, a tract of about 10 miles is passed before the line is again met with in any state of forwardness. It begins to take form opposite Corrour Shooting Lodge and a mile of so further north it is nearly ready for the metals.'

Faced with a boggy morass, Lucas & Aird could have tipped hundreds of tonnes of rock into the moor and hoped for the best but in the end the company decided to adopt a method employed by George Stephenson in the 1820s to cross Chat Moss, on the Liverpool and Manchester Railway. Effectively, the line would be floated across the moor.

A frontier town of sorts sprung up around the site of Rannoch Station and huts were erected at Gorton. Access was improved by extending the road west from Bridge of Gaur and a standard gauge light railway was constructed to ferry materials and men around.

Lucas & Aird built a stone boarding house for its engineers and surveyors and huts erected by both the contractors and enterprising individuals provided accommodation for navvies employed in the final push. There was also a four-bed hospital and a shop.

Access to clean drinking water, however, was a problem at Rannoch and Gorton, the water coming off the moor considered too dangerous to consume. Hut keepers at Rannoch could sell beer, which the navvies drank with their breakfast, lunch and dinner. A similar dispensation was not given up the line at Gorton where two female hut keepers quickly fell foul of the law when they dispensed alcohol to their resident navvies.

Sarah McLeod, wife of a hut keeper at Gorton, and Catherine Bryne, also of Gorton, were both called to the dock at Oban Sheriff Court in July 1894 charged with selling 'excisable liquors' without a licence the previous month.

The court heard that due to the bad water at Gorton, it was the custom of the 30 or so men staying at McLeod's hut to take beer with their meals, and it was a similar situation at the hut run by Bryne and her husband. Custom it may have been but legal it was not and both women were fined £2 and ordered to pay expenses.

While Lucas & Aird was able to deploy a sizeable workforce to Rannoch Moor progress was hampered by delays in getting materials to the site. The *Glasgow Herald* noted that were it not for this, the line could have progressed across the moor at the rate of four miles a month.

Construction was overseen by engineer John Blue, who spent his entire working life with Lucas & Aird.

Born in Campbeltown in March 1850, he joined the company in 1875 and represented the firm on many important contracts, including railways in the UK and the Tilbury Docks in London. He also worked on the Aswan Dam, on the River Nile in Egypt, and, prior to his death in March 1909, Avonmouth Docks, Bristol.

To cross Rannoch Moor, where the soft peat varied in depth from a foot to 16 feet, he instructed the navvies to dig trenches and cross-drains and layers of turf and brushwood were laid into these.

Often, the material simply disappeared into the mire, but the process was repeated until a reasonably solid base eventually took shape. On to this, thousands of tonnes of soil, ash and stone were tipped.

'Taking advantage of spells of dry weather recently experienced, long stretches of the route through the moor, which is peat moss from end to end, have been deeply channelled on each side to drain off the water,' the *Glasgow Herald* noted.

'Where the natural courses show that the water flows across the line of the railway, box-drains and culverts are being put down. It is the getting of the boxes and the cement for the culverts on to the moor that is the difficulty at present. The boxes are lumbering affairs of heavy timber which, with their coatings of tar, will remain serviceable pretty nearly as long as the railway lasts.'

Progress was depressingly slow and money tight and the bog swallowed the last of the railway company's funds. There was a very real risk that all involved would lose their shirts on the Moor of Rannoch.

Thankfully, railway company director James Renton, a stockbroker, came to the rescue, providing a generous bridging loan that enabled construction to proceed. His investment may have been prompted in part by the fact his daughter Alicia Ellen was married to John Aird, son of contractor John Aird.

In his honour, navvies sculpted his relief portrait into a boulder – the Renton Stone – which they manhandled into place at the northern end of Rannoch Station platform. It remains there to this day. He was also given the accolade of driving home the final spike in September 1893.

Renton Stone

With the railway slowly progressing across the moor, the location of stations had yet to be finalised and, as the *Glasgow Herald* reported, in 1892 Corrour did not appear to feature in the railway company's plans.

The newspaper added: 'On the moor there are to be two stopping places for the benefit of sportsmen, one at Achallader, opposite the Marquis of Breadalbane's shooting lodge, Forest House, and another at the River Gaur, the well-known trouting river which connects Loch Laidon and Loch

Rannoch. The plans do not show any more stations from this point until Glen Spean is reached.'

Achallader would become Gorton while the halt on the River Gaur would open to passengers as Rannoch Station. A third site at Luibruaridh would later be pencilled in as a passing place and private halt to serve Corrour Estate. It would be named Corrour.

The joy of conquering the morass and, in the process, completing the line, was tainted by the knowledge that the last deaths to occur during construction happened here.

On April 16, 1894, navvy John Campbell was, according to rather starkly written Press reports, 'blown to pieces' during blasting work, while on June 18, labourer John Dolan was run over by a locomotive. He lost both legs and died.

Later in the month, 11-year-old Peter Thomson, son of a hutkeeper on Rannoch Moor, was knocked unconscious after being struck in the face while attempting to sprag a wagon (the placing of a steel bar through the spokes of a wheel to prevent the wagon rolling downhill). He was conveyed to hospital in Fort William and made a full recovery.

In a final unfortunate incident, on July 30, 1894, a man named John Smith was attacked at Rannoch Station by an angry navvy who had just been dismissed. Threatening 'to do' for somebody, the labourer passed Smith, took a knife from his pocket, and stabbed him. He was promptly apprehended and handed over to the police while Smith survived the unprovoked assault.

On a lighter note, the navvies were invited to compete in athletic games organised at Loch Treig. They, however, were at an immediate disadvantage for while the prizes on offers to locals were 10s, 7s 6d, and 5s for first, second and third place respectively, the navvies were awarded less for their race wins – 7s, 5s and 3s. If nothing else, it was a little extra beer money.

Track laid there were numerous jobs to complete and problems to solve before the line could be inspected ahead of its ceremonial opening on August 11, 1894.

Revenue earning passenger and freight trains had already been running for four days when the Marchioness of Tweeddale, wife of West Highland Railway chairman William Hay, Marquis of Tweeddale, turned a golden key to admit the approaching North British train into the gleaming new Fort William terminus.

From the outset, the board had tried its utmost to keep costs to a minimum, using cheaper, lighter rails, recycling rock and soil excavated from cuttings into embankments and bridge piers and taking the least expensive line through the countryside wherever possible.

Lucas & Aird's original tender was £393,638, although they negotiated an additional £10,000 bonus on completion during the crisis talks of October 1891. In the summer of 1892 the cost was put at £600,000 (£6000 a mile) but in the end, it is estimated that the West Highland Railway cost £11,000 a mile to construct, just over £1 million in total, suggesting its saviour James Renton had very deep pockets indeed.

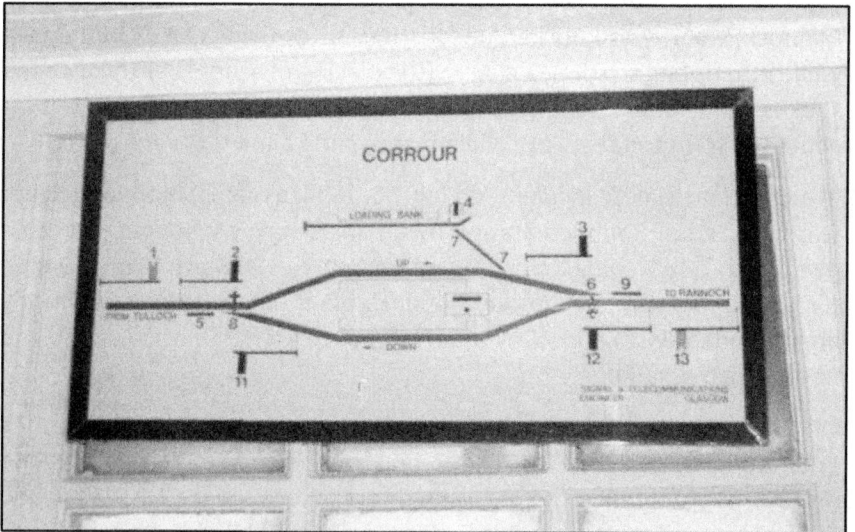

Signalling diagram within the box at Corrour

4

Moorland Outposts

Completed thanks to great engineering ingenuity but at significant financial cost and loss of life through both accidents and illness, the railway over Rannoch Moor served three remote outposts. Two – Corrour and Gorton – were officially private halts, equipped with signal boxes, passing loops and sidings, while the third – Rannoch – was a fully-fledged station.

As the line between Craigendoran and Fort William was single-track throughout, it relied on these passing loops, installed at regular intervals, to allow trains travelling in opposite directions to safely negotiate one another.

An electric single line token system was installed by Saxby & Farmer who, from the 1860s to the 1880s, were the dominant force in railway signalling equipment manufacture. Partner John Saxby is credited with inventing the interlocking system of points and signals, a major advance in railway safety.

To enter a section of track, approaching locomotive drivers had to hand over the token for the previous one and pick up the token for the next. This was often done at some speed and the tokens were contained within leather pouches with a long loop handle to enable then to be passed to and from a moving train. Stations and signal boxes were linked by a telephone line, the overhead cables carried on poles running alongside the track.

Travelling north from Glasgow, nine miles down the line from Bridge of Orchy, trains climbed steadily to Gorton – known as Gorton Siding or Gorton Crossing – where the ancient Rannoch drove road, a route well used by farmers moving cattle and sheep from the West Highlands and islands to markets in Crieff and Perth, crossed the railway.

Located on the southern fringes of the moor at an elevation of around 1100 feet, the halt – originally called Gortan to avoid confusion with Gorton Station in Manchester – consisted of a passing loop with an island platform and siding.

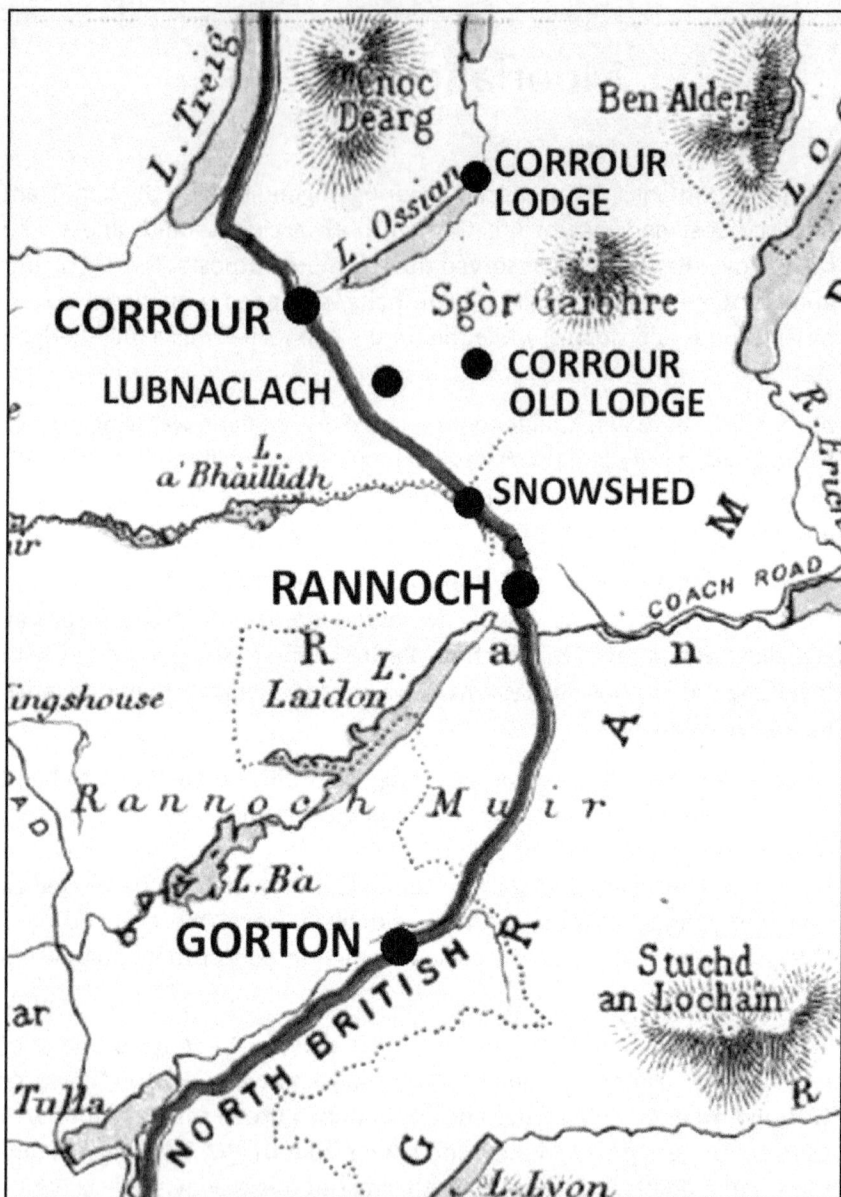

A map showing the locations of Gorton, Rannoch and Corrour, along with other
notable features including Lubnaclach, Cruach Rock Snowshed, Corrour Lodge
and the structure it replaced, Old Corrour Lodge

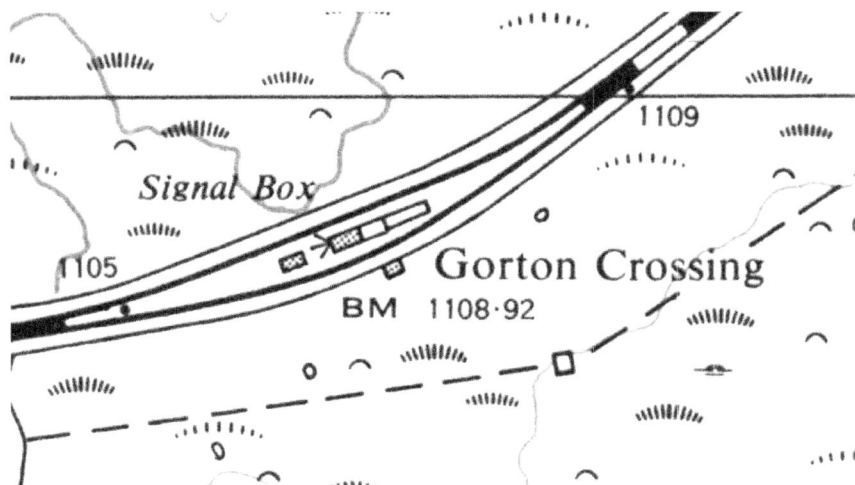

Gorton Crossing

The architecture differed at Gorton from the normal West Highland Railway station. Instead of the traditional platform-mounted chalet-style building and neighbouring signal box, it had a tall signal box – often referred to as the 'watchtower' – attached to which there was a rectangular single-storey building in which the signalman lived. It was constructed of rendered brick with banded dressings and a grey slate roof with terracotta ridge tiles. There were only two other such buildings on the line, one at Corrour and one at Glen Douglas.

There was no paved road to the station, only the unsurfaced drove route, and all supplies arrived by train. The nearest neighbours were the shepherd's cottage at Gorton, a mile or so to the west, where members of McAlpine's ill-fated expedition took refuge during their moorland traverse, and track workers' cottages up and down the line.

From the outset, the Marquis of Breadalbane, on whose land the halt stood, was adamant that Gorton would not serve the public. In September 1896 he petitioned the Court of Session in Edinburgh, demanding that the West Highland Railway Company do not maintain any station for passengers or goods either at Gorton or elsewhere on the line where it crossed his land north of Bridge of Orchy, without his consent in writing.

The Marquis had noted that following the opening of the line two years earlier, the company had not only constructed a platform at Gorton and erected nameboards but had also been in the habit of stopping trains there to set down passengers and parcels, violating the agreement that allowed the line to proceed across his estate. In the original Act, the Marquis had only consented to stations at Tyndrum and Bridge of Orchy.

He won his argument and Gorton was relegated to private halt, the signalman – or indeed any railway worker – threatened with instant dismissal if they allowed members of the public to board or alight there.

Gorton nameboard, 1964

From Gorton, the train puffed north – never exceeding 25mph – across the moor for seven lonely miles, climbing steadily through a largely featureless landscape of exposed grey granite and dark peat bog peppered with dead tree roots and inky black pools, before dipping to cross the five-girder Gaur Viaduct beyond which Rannoch Station appeared.

Rannoch Station (Author)

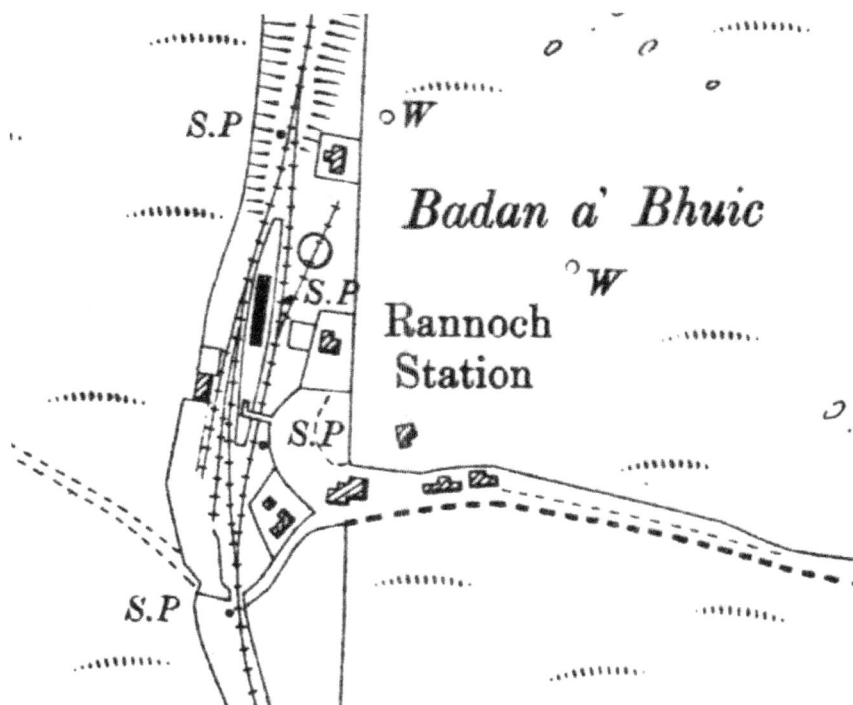

Rannoch Station, 1897

Perched at an elevation of 977 feet, Rannoch was officially the only public station on the moor. A level crossing at the southern end of the island platform allowed a drove road from King's House, in Glen Coe, to cross the track before proceeding east by Loch Rannoch.

The platform itself stood in the centre of the passing loop and there were two sidings to the west, a freight platform and small wooden goods shed next to which a corrugated iron shed stood for many years. A spur to the east served a turntable, which was later removed. There was also a water column fed by a tank hewn from a repurposed locomotive boiler.

The station building was one of the iconic Swiss chalet-style structures typical of the line, designed by Scottish architect James Miller. Long and low with a deep-swept roof and overhanging veranda on all sides, the building featured a red brick base mounted on a stone plinth, timber-framed above with scalloped shingle (imported from Switzerland) walling, windows and tall brick chimney stacks.

Inside, from one end to the other, the building housed men's urinals, two toilet cubicles (one for men, one for ladies), a ladies' room, a general waiting room, access passage, booking office, stationmaster's office, porter's room and a room for storing luggage and parcels. A sub-Post Office opened on July 15, 1895, and Rannoch Station became a Post Town in August 1905.

Adjacent but separate to the station building was a small signal cabin of similar architectural style housing 17 levers and, beyond this, at the northern end of the platform, the Renton Stone.

Access to the station was by road from Bridge of Gaur and access to the platform itself, from a parking area cum turning circle at the end of the road, where there was a telephone box and shelter, was by footbridge. The original wooden crossing was replaced by a lattice metal structure moved south from Corrour in the 1980s.

In common with all the stations and passing loops on the railway, trees were planted around the platforms to afford some shelter from the prevailing wind.

The ascent from Rannoch Station to Corrour – the highest point on the moor – was one of the most arduous locomotive, driver and fireman faced on the journey from Glasgow to Fort William.

Sliding out of Rannoch Station, the train crossed the curving lattice girder Rannoch Viaduct, now one of the most photographed points on the route, before climbing over the moor, skirting between high snow fences designed to keep winter drifts off the exposed section of track.

The rails enter Cruach Rock Snowshed, a 205-yard-long cutting covered with a roof of corrugated iron, before emerging on to open moorland once again, passing below Old Corrour Lodge, perched on the slopes of Carn Dearg to the east, and the cottages of Lubnaclach, where surfacemen tasked with maintaining this section of the route were billeted.

A little over a mile further on, the train slowed for Corrour, the highest station on the West Highland Railway and, at 1,340 feet (408 metres) above sea level, the highest mainline station in Britain.

The passing loops at Corrour and Gorton shared the same signal box/station building design

Ordnance Survey maps from the last years of the 19th century refer to the station as Corrour Siding and show a passing loop, island platform with buildings, a single siding to the east and a footbridge. Rival cartographers Bartholomew christened the site Corrour Passing Place. In later years it would appear on maps as Corrour Station.

A 13-lever signal box, identical to the structure at Gorton, was constructed at the southern end of the platform. Attached to this, on the platform end, was the same rectangular single-storey building in which the signalman lived.

Over the years various extensions were added including a brick-built hut on the other side of the box and a lean-to access corridor on the west side of the building to provide the signalman with some additional protection from the elements.

Although Corrour was constructed as a private halt, landowner Sir John M Stirling Maxwell was an ardent supporter of the new railway and did not pursue the same hard line as the Marquis of Breadalbane did at Gorton, allowing the NBR to offload both people and parcels from the outset. A wooden waiting room with slate roof stood on the platform itself.

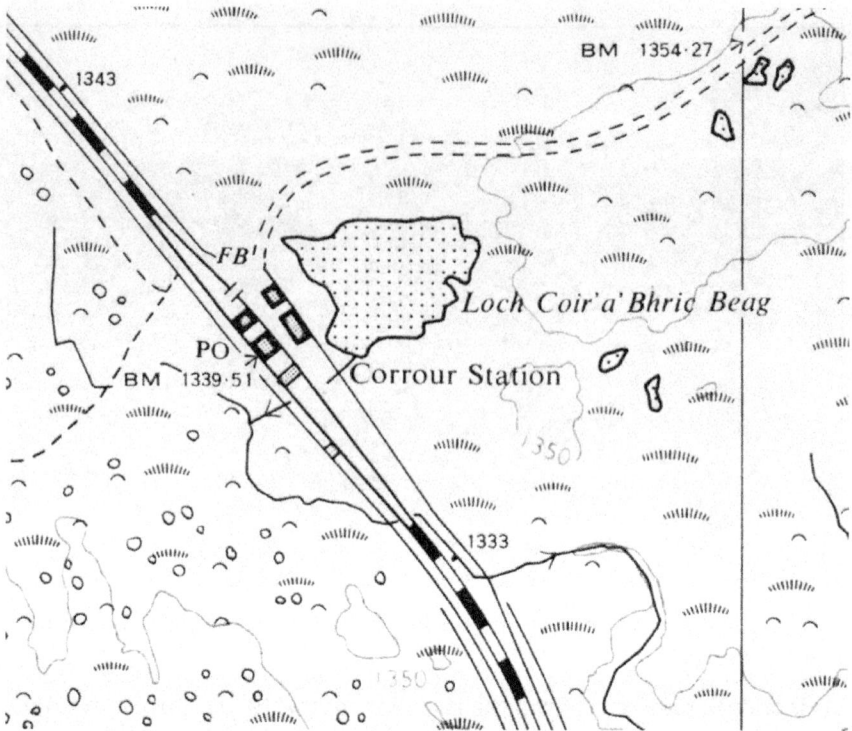

Corrour Station, 1968

On December 15, 1896, a sub-Post Office opened within the station and a Telegraph Office was added on August 16, 1898. To make space for these and to accommodate the increased amount of administrative work relating to estate traffic, particularly during the shooting seasons, the railway company built a single-storey rendered brick cottage for the signalman and his family opposite the signal box and created offices within their former home. In 1905 Corrour Station was designated as a Post Town by the Post Office. A brick-built shed was also erected adjacent to the cottage.

Following his purchase of the estate in 1891, Sir John commissioned the construction of a new lodge at the eastern end of Loch Ossian to replace the one on

the western shoulder of Carn Dearg, declaring it both too small and too far from the heart of estate activities, and he built new cottages for estate workers, improved the estate's tracks and planted forests. At the time the estate derived much of its income from grouse shooting and fishing. Sheep farming was in decline and over time the rough grazing land he acquired from the previous owner, Colonel Sir George Gustavus Walker, was replaced with deer forests.

Among the new cottages was one for estate keeper John MacDonald, who ran the Beinn Bhreac beat. He lived out on the moor at Caim, west of the railway, but in 1893 a house was built for him at a less remote spot at Luibnaclach, on the eastern side of the track.

Loch Ossian (Author)

As summer passenger traffic and freight increased over the line, plans were mooted to create eight new passing loops and signal boxes, splitting some of the longer sections. One was proposed for Caim, midway between Rannoch Station and Corrour, creating a fourth outpost on Rannoch Moor. The idea, however, never came to fruition.

The railway ferried both the materials and workforce in for Sir John's new lodge, which was complete by 1899. It also imported a steam yacht, *Cailleach*, which was built by Matthew Paul & Company Ltd of Dumbarton in 1902 to ply the waters of Loch Ossian, transporting sporting guests arriving by train from a wooden boathouse at the western end of the water to a pier below his great granite edifice. At the time there was no road between the station and lodge although one was laid along the north side of the loch in 1911.

The lodge boasted 14 bedrooms, three bathrooms and four public rooms, drawing and dining rooms panelled with pine and adorned with oil paintings, family crests and stags' heads. Equally lavish were the formal gardens, flanked by a chapel, game larder, laundry, schoolhouse and cottages and surrounded by new woodland. In 1939 plans were drawn up to add a recreation hall but the outbreak of the Second World War scuppered them.

Both the lodge and outbuildings had electricity, generated by a small dam on the Corrie Creagach burn. It was all a far cry from the rather more primitive accommodation offered by the Old Lodge which was repurposed briefly as lodgings for servants before being abandoned. The 1901 census shows it uninhabited at that time. The roof was later removed and the building fell into ruin. The map above shows the layout of the old lodge in 1870.

Still flanked by snow fences, beyond Corrour Siding, the railway passed over Corrour Summit (1350 feet/411 metres), where signs either side of the track recorded the elevation. These were originally wooden but were later replaced by a more robust iron framework. Thereafter, the line left the moor and descended by Loch Treig to the next station, Tulloch (renamed from Inverlair on January 1, 1895), 10 miles to the north, beyond which it swung west for the last 18 miles of the journey to Fort William.

While Corrour station was and still is the highest mainline station in Britain, Corrour Summit was not the highest point on the railway network. That accolade goes to Druimuachdar summit (1,484 feet/452 metres), in the Pass of Drumochter, on the Perth to Inverness Highland Railway, although it was held by the Leadhills & Wanlockhead Light Railway branch (1498ft/456m) until its closure in 1939.

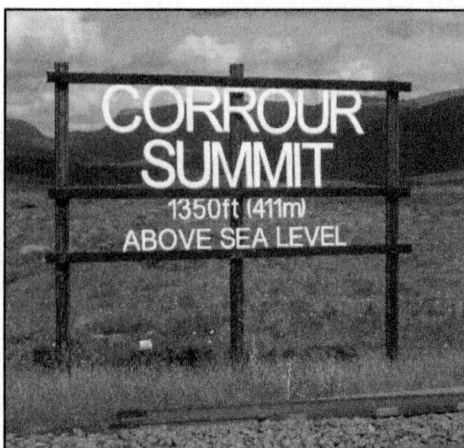

The Corrour Summit sign as it stands today (Author)

5

Early Days

In a bid to recoup their investment as quickly as possible and find out exactly what levels of traffic they could expect over the West Highland Railway, the board of directors opened the line to passenger and freight services before it was ready.

The first passenger train carrying fare-paying customers left the platform at Fort William shortly after 6am on August 7, 1894. Extra coaches were laid on to deal with the demand, much of it from inquisitive locals, many of whom had never travelled by train. Most only went as far as Spean Bridge or Roy Bridge, returning on the 7.30am train from Glasgow which got into Fort William just after noon.

While the track and signalling system were in place and had been approved by the Board of Trade, many of the stations had not been finished, sidings had not been laid, and insufficient accommodation had been provided for the stationmasters, signalmen and surfacemen who were to work the line.

The NBR had been successful in its recruitment drive, drawing staff from all corners of its network. Many were keen to swap town and city postings for a new life in the West Highlands and the company was keen to attract family men to Corrour, Rannoch and Gorton, knowing they were more likely to stay longer.

Pay rates were lower but, as the company pointed out, to be a clerk in Glasgow on £80 a year was to be a nobody, while to be a stationmaster on £60 per annum, with a free house and coal, was to be a notable figure in a small community.

While stationmasters on the West Highland Railway, who undoubtedly became respected figures, well known in their local areas, pocketed £60 a year, signalmen could expect a wage of 16s a week, although they were expected to pay rent for their accommodation. However, as part of the drive to encourage staff to the remoter corners of the network, the housing provided by the NBR was of a higher standard on the West Highland Railway than it was at the company's other stations.

Railway cottages at Rannoch Station (Gordon Hatton)

Better they may have been, but the promised perks were still some way off. At Rannoch Station, for instance, a cottage for the stationmaster and his wife was not yet ready and the couple were forced to make do with a damp and draughty hut abandoned by the navvies. Living in such conditions, the stationmaster's wife soon fell ill and died, and her distraught husband is said to have 'completely lost heart' thereafter. His career on the West Highland Railway was sadly short.

The porter and signalman at Rannoch found themselves billeted to the public waiting room, while the signalman at Tulloch and the signalman, his wife and daughter at Spean Bridge were forced to live in the station waiting rooms.

Surfacemen along the line often found themselves housed in old wooden huts repurposed from the work camps, rather than the cottages promised. It was a year, sometimes two, before all the new housing was ready.

At Corrour, while the accommodation appears to have been in order, the siding had yet to be laid and gates were not installed in lineside fences, forcing gamekeepers and their ponies to make lengthy detours to cross the track.

The siding remained unfinished until April 1895 and in the meantime materials for construction of the lodge and cottages, tree saplings and baby trout were unloaded direct from the mainline, or offloaded at Tulloch or Rannoch, neither particularly convenient for Sir John.

At Corrour Siding, the railway initially employed a single man who looked after all aspects of railway life. In 1902, an article in *The Railway Magazine*, entitled 'A Many-Sided Man', summed up his various responsibilities.

Home to a 'many-sided man', the signal box at Corrour

It stated: 'For a lonely and many-sided job there are not many to compare with that of the station agent at Corrour Station who acts as stationmaster, signalman, porter, goods clerk, booking clerk, parcel clerk, telegraph clerk, postmaster and postal telegraph clerk.

'He is perched on the top of a hill in Inverness-shire, 1350ft above sea-level, two miles from the nearest neighbour, ten miles from a school, twenty-eight miles from a doctor, baker, butcher, shoemaker or tailor, yet the number of letters that pass through his hands is wonderful.

'For two months over 6000 letters and parcels came by post, 800 parcels by train and, in addition, 600 postal telegrams were received.

This Macintyre's Series postcard shows, on the left, the house constructed for the signalman and his family at Corrour and, in front of it, Annie MacNicol holding her young son

'If he wishes to send a letter to his nearest brother-in-trade, seven miles distant, it has to cover 185 miles before it reaches him.

'In the morning he finds the grouse sitting on the top of his house, and the window-sill, and often enough the red deer and mountain hare eat his kale.'

With the opening of the Post Office and Telegram Office and increased passenger traffic, the workload was generally shared by a married couple.

Census records show that in 1901 the Corrour signalman was 24-year-old Alexander Macdiarmid, a single man from Killin. The 1911 census reveals that the signalman was 34-year-old Archibald MacNicol, from Kilmallie, Fort William, who lived at the station with his wife Annie (30), from Edinburgh, and their one-year-old son James. The couple had come to Corrour in 1908 and would remain there until 1941.

At Rannoch Station, there was a larger staff. In addition to the signalman, there was a stationmaster and porter. The 1901 census shows that 34-year-old stationmaster Robert Hogarth lived with his wife Isabella (35) and their three-year-old son William in their own cottage while 37-year-old signalman Henry Purves, who was married, boarded in the home of Angus McRae (45), a surfaceman, who lived with his wife Margaret (44) and son Donald (20), who was employed as station porter.

It was a busy household for in addition to their two younger sons, four-year-old Jonathan and one-year-old Thomas, the McRae's provided bed and board for track workers John McLean (35), John McDonald (43) and Jonathan McLeod (28).

In a neighbouring cottage, surfaceman Allan Lamond (61) lived with his wife Rachel (57), and they had two lodgers, both single men employed as surfacemen.

With the opening of the railway, general merchant and carrier Duncan Cameron, who ran a well-established store in Kinloch Rannoch, expanded to Rannoch Station, erecting a wooden-framed corrugated iron general merchants shop, selling groceries, provisions, wines and spirits.

Cameron also acquired the lodging house constructed for engineers and surveyors in which he opened a public bar. He died in 1906 at which point his businesses interests in both Kinloch Rannoch and Rannoch Station were put up for sale.

They were acquired by his son Archibald Cameron who appears to have moved the shop into the former lodging house and, adding letting bedrooms and refreshment rooms, opened the Rannoch Station Hotel in the 1930s.

There was also a railway company coal depot and a motor garage run by A McKerchar, of Aberfeldy, although by the 1930s the latter was sitting vacant.

Between Rannoch Station and Corrour, the men employed to maintain the track over the moor were billeted at Lubnaclach where there was a substantial three-bedroomed stone house, home to a gamekeeper employed by Corrour Estate, and two cottages occupied by the railmen. In 1901, foreman Allan Cameron (29) lived with his wife, two young sons and two lodgers, both surfacemen, in one of them, while the other provided lodgings for four surfacemen, all single men.

In 1911, foreman John MacKinnon (22) and his 33-year-old wife Mary lived with three lodgers, Angus Dingwall (25), Duncan Beaton (23) and James Munro (21), while next door 31-year-old John MacLellan and his 27-year-old wife Sarah had two boarders, Angus MacLellan and John Morrison, both 19.

The former gamekeeper's cottage at Lubnaclach (Author)

While nine people occupied the two railway cottages, gamekeeper Donald Lawrie (28) had only housekeeper Mary McIntyre (55) for company in the larger estate house.

Lubnaclach, 1904

Gortan – renamed Gorton on May 1, 1926 by the LNER – was without a doubt the most isolated of the Rannoch Moor stations, the result of its location and the fact that for much of its life passengers were not allowed to board or disembark here.

Between the 1930s and 1960s, the carriage school mounted on the platform injected some life into the place during the week and there was plenty of activity during the shooting seasons, when the Marquis of Breadalbane allowed the NBR and later the LNER to disgorge his guests at the station.

Southbound LNER J39 locomotive 4950 hauls a freight train across Rannoch Moor between Rannoch Station and Gorton Crossing. Built in Darlington, the engine entered service in 1938 and continued into the British Railways era, ending her days at Inverurie Works in 1963 (Colour-Rail)

In terms of neighbours, there was the shepherd's cottage across the glen and surfacemen's cottages at milepost 52½, to the south, and milepost 58, to the north.

In common with Corrour and Rannoch, there was no electricity here. Coal heated the signalman's home and box and paraffin powered his lamps. In later years bottled gas and batteries found their way to this remote outpost.

Drinking water was delivered by train along with milk and food while mail was bagged at Bridge of Orchy and sent down the line.

The NBR, LNER and latterly British Railways struggled at times to staff the box, a situation that hastened its eventual demise in the 1960s.

In 1915 the signalman was Donald Grieve and through the 1920s his son Thomas Grieve held the position. Harold Hasty was signalman throughout the 1930s and in 1940 the role was taken by Lachlan McLeod.

In addition to the permanent staff, the NBR and its successors employed relief signalmen who were deployed to cover days off, holidays, sickness and other absences. It is said that in the early days they travelled equipped with a camp bed, kettle and snare, all they needed to survive on the moor.

While the relief men could find themselves bedding down in the signal box, more often than not, if the family was away on holiday, they were given the run of the railway cottage. While more comfortable, this could entail a lengthy list of extra duties, such as looking after the animals and ensuring the vegetable patch was kept watered.

With staff living in sub-standard conditions and the infrastructure incomplete, the railwaymen of Rannoch Moor successfully dealt with a busy end to the summer of 1894.

A hectic first shooting season followed with extra workers drafted in to cope with the arrival of the sporting parties and the dispatch of the red deer and grouse bagged on the estates flanking the track.

The NBR prided itself on the punctuality of its passenger services on the West Highland Railway, in contrast to the rival Callander & Oban Railway where it was well known and often remarked upon that trains frequently ran up to an hour and a half late.

Keeping to schedule was not always easy and 1894 ended with a real sting in the tail that made it, at times, completely impossible. The West Highland Railway's first winter was, according to records of the time, the worst of the 19th century.

Fierce snowstorms in December were swiftly followed by the Great Frost of January 1895, temperatures plummeting as low as minus 18 degrees centigrade in northern Scotland.

The season started benignly enough with a mild and wet November and, had this continued, the West Highland Railway would have been fine. However, at the end of December, colder air funnelled across the country, bringing with it heavy and sustained snowfall.

McAlpine, Forman, Harrison et al knew from first-hand experience what winter on Rannoch Moor was like and ultimately it did not disappoint.

The first snows of December cause few problems and disruption to services was minimal. Then, on January 8, the *Daily Telegraph and Courier* reported that 40 miles of the West Highland Railway were impassable due to blizzarding snow. Telegraph lines were blown down and timetables went out the window. Trains left Glasgow on time but only made it as far as Crianlarich.

Stranded snow plough at Corrour, 1894 (John Alsop)

'The early goods and passenger trains had to return to Glasgow on reaching Crianlarich, and a snowplough which left the Fort William end in the morning got buried in a fifteen-feet drift at Tulloch,' the newspaper stated.

'A snowplough left Crianlarich to clear the track northwards but up to a late hour last night nothing had been heard of it.'

With ploughs lost on the line, *The Scotsman* noted on January 8 that the 'chief feature of the storm appears to have been not so much the amount of the snowfall as the fierceness of the wind which accompanied it and distributed the snow in deep drifts among the valleys.'

This cut communications between Glasgow, Oban and Fort William and while the line to Oban escaped relatively unscathed, booking clerks in

Glasgow stopped selling tickets to Fort William while the snowploughs and gangs of railwaymen with picks and shovels toiled day and night to dig the tracks out.

The following day, the *Glasgow Herald* reported that snowdrifts were still blocking cuttings between Rannoch and Tulloch stations and for the next few days the line remained closed over Rannoch Moor as falling snow and drifts swiftly undid the work of both the men and the ploughs.

While some progress might have been made, the *Dundee Evening Telegraph* reported in January 16 that the line between Gorton and Corrour remained closed. If it was any consolation to the West Highland Railway directors, who were paying out money to clear the line but not taking in a penny in revenue, the Highland Railway to Inverness was facing an equally difficult time.

Eventually the railway was cleared, timetabled passenger services returned to normal and freight trains were resumed between Glasgow and Fort William. However, it was but a brief respite for in early February another wave of blizzards hit the West Highland Railway, snow and freezing temperatures once again closing the northern section of the line on February 6.

The *London Evening Standard* reported on February 8 that a ballast train brought to a halt near Rannoch Station two days earlier remained embedded in the snow.

Motive Power

Two classes of steam engine were designed and built specifically for use on the West Highland Railway, even though it had no workshops of its own.

The first was the NBR Class N, a 4-4-0 created by Matthew Holmes and built at the company's Cowlairs works in Springburn, Glasgow. A dozen were made in 1894 and another eight were added to the fleet two years later.

Known as the West Highland bogies, they performed poorly and, despite a rebuild, no more were built. All but one had been scrapped by 1924.

The later K Class, a 4-4-0 designed by William Reid, proved to be more capable. Known as the Glen Class (engines were named after Highland glens) the loco was deployed to the route in 1913 and, for over two decades, it was the main workhorse, hauling most of the passenger services and continuing in service until the withdrawal of steam.

'It is feared that the workmen aboard her will have some difficulties in obtaining provisions,' the writer continued. 'An engine and snowplough started from Fort William but left the rails near Loch Treig. The driver and fireman had a narrow escape, for another foot beyond the engine would have been precipitated down a gully 50-feet deep.'

Once again, snow ploughs and teams of men were dispatched to clear the tracks and, while the line was briefly re-opened, renewed bad weather blocked it again on February 14 and it remained closed for the next four days

On February 20, a thaw finally set in and the line re-opened across Rannoch Moor, allowing trains to run unimpeded between Craigendoran and Fort William. It would not be the last harsh winter the railwaymen and women of the moor would face but, lesson learned, swift steps were taken to try and lessen the impact of drifting snow on the railway in the future.

The sides of the cutting at Cruach Rock, north of Rannoch Station, were raised and a corrugated iron roof was slung across the top, creating the Cruach Rock Snowshed, a unique structure in Britain, while sturdy snow fences constructed from rails and sleepers were embedded in the moor where the drifting had been particularly severe.

Gangers dig out the track at Corrour Station

Amid all this winter chaos, the West Highland Railway Company held its half-yearly general meeting in Edinburgh where its chairman, the Marquis of Tweeddale, reported that the line had so far proved 'very successful', with traffic exceeding expectations.

However, on August 6, 1896, in the House of Parliament, NBR general manager, Mr John Conacher, revealed a rather less rosy picture. He had been called to give evidence at a Select Committee hearing considering the Invergarry & Fort Augustus Railway Bill.

Following the opening of the West Highland Railway, plans were drawn up to extend the line west from Fort William to Mallaig, to capitalise on fish traffic, and construct a branch north through the Great Glen from a junction near Spean Bridge to Fort Augustus, at the southern end of Loch Ness, the next step in a campaign to eventually reach Inverness and snatch traffic from the Highland Railway Company.

Ministers were also considering the West Highland Railway Guarantee Bill which would provide a subsidy for the line to Mallaig, perhaps not surprising as the government had backed the project as a way of tackling high levels of poverty in the region by providing the population with better transport links.

Conacher revealed that the earnings of the West Highland Railway had, to date, been 'most disappointing'. The committee heard that in the first year of operation income had only amounted to about five guineas per mile, per week, less than half what was expected.

On this basis, government subsidy would amount to £2000 a year for at least 30 years, although the Chancellor of the Exchequer, Sir Michael Hicks Beach, was optimistic income would increase and the final bill would be much less.

Despite Conacher's comments, the extension to Mallaig was approved, the line opening on April 1, 1901. McAlpine's moorland misadventure paid off as his company was awarded the construction contract and he was able to demonstrate to the world just what he could do with concrete.

While the line to Mallaig delivered healthy fish freight on to the main artery of the West Highland Railway, the branch to Fort Augustus, which opened in 1903, only proved useful during the Second World War and it closed in 1946. The NBR never reached Inverness.

In addition to the Mallaig extension and Fort Augustus branch, other schemes were mooted to take advantage of the new West Highland Railway.

In 1895 well-attended public meetings were held in Tulloch and Newtonmore, a station on the rival Highland Railway, to garner support for a line following in part the stagecoach route that ran through Laggan from Tulloch to Kingussie. Prior to the opening of the West Highland Railway this had been the easiest if not most direct way to reach Fort William and Lochaber.

In July 1903, the *Dundee Courier* reported on another set of plans, these ones drawn up by an unnamed engineer, to construct a railway from Rannoch Station to Dalwhinnie, via Loch Ericht. While the Tulloch to Newtonmore proposal resurfaced in the 1970s, threatening the future of the line across Rannoch Moor, the Loch Ericht idea sank without trace.

There is no doubt that poverty was rife in the western Highlands and islands and in the years leading up to the construction of the line, there was much debate on what could be done to assist the crofting and fishing communities without resorting to hand-outs. Improving transport links was one way to help people help themselves.

The Napier Report, a study of the living conditions of crofters and farmers in the Highlands and islands, commissioned by the government in 1883, concluded that better communication by post, telegraph, roads, steam vessels and railways was needed.

There were not many people living on Rannoch Moor at the time, but it was not too difficult to find an example of crushing poverty here, one that highlighted the plight of the ordinary West Highland family struggling to survive in this cut-off corner of the country during the latter part of the 19th century.

Almost a decade before the railway opened, in August 1885, the *Aberdeen Free Press* published an account of a walk across Scotland by an unnamed correspondent which, in its course, crossed Rannoch Moor.

At that time, the carriage road, which approached along the northern shore of Loch Rannoch, ended at a farm at Dunan and a rough cart track progressed west to Loch Eigheach where it swung north, climbing to the old Corrour Lodge before progressing north by Loch Ossian and Loch Treig to Glen Spean.

A family of tinkers and their camp in Highland Perthshire (National Library of Scotland)

The author, however, continued west, following the line of the modern day B846 – then just a rough path over the peat moss – to a corrugated iron-roofed shepherd's cottage at Doire na h-Innes (now a ruin) and on by the River Gaur across what would become the site of Rannoch Station to Dubh Lochan and Loch Laidon.

Along the way he and his companion encountered a gypsy encampment, a common enough site in Highland Perthshire in those days, and a family of travelling merchants walking to Lochaber, the rag-clad children already footsore with another 20 miles of hiking ahead of them.

Passing a cottage at Cruach, by Dubh Lochan, the newspaper's correspondent journeyed on along the lochside, pausing for an hour or so at Tigh na Cruaiche, isolated home to a poor shepherd and his family.

'At Cruach we have reached the acme of desolation and dismalness of the famed Moor of Rannoch, ocean of blackness and bogs and, without exception, the dreariest part of Scotland. Except the solitary hut of Cruach there is nothing visible around but the rough gloomy moor, with gloomy a loch, Loch Laidon, in its centre,' he wrote.

'At Tigh na Cruaiche during certain seasons of the year not a single person will pass for fortnights, we were told, at a time.'

The original waiting room at Corrour

The cottage itself, he noted, was a 'very wretched structure', loosely built of stones and with a tumble-down thatched roof, indiscernible from the moorland around it, a wide-open chimney, and a broken window.

'Such an erection as this hut would not be tolerated by law in a large town as fit habitation for human beings. But on the Moor of Rannoch, and such regions, the law takes no concern with such matters and allows a family to be raised in a hovel that most men would not allow dogs to live in.'

While impressed by the family struggling to survive here, the reporter was less enamoured with the landowner who had allowed their home to deteriorate to such a state of squalor.

With one end of the building devoted to cattle and poultry, the shepherd and his 10 children occupied the other. With no school in the vicinity, lessons were provided by a visiting teacher.

'It seemed a heartless affair altogether, the only redeeming feature being the shepherd himself and the children, including a well-bred daughter,' the writer noted as he and his friend set off into the wilds, revived by the generosity of the poor family who supplied them with a meal of milk and bread.

It is not known what became of the family or whether they benefitted from the coming of the railway 10 years later. The decline in sheep farming at the end of the 19th century may have brought their hardships to an end, or posed new challenges. Their home stands today as a ruin on the moor, their struggles reason enough to subsidise better transport links.

During the early years of operation, the NBR staff on Rannoch Moor were by no means inundated with trains.

Headed by two British Railways Standard Class 5 locos, a passenger service from Fort William draws into Corrour, 1961 (R B Parr)

At Gorton, Rannoch and Corrour, the day started with the passing of the daily early morning freight train from Glasgow, a service known as The Ghost. It dictated the time at which the signalman had to be up and out of bed and at his post, ready to exchange tokens. Often this was as early as 4am.

Thereafter, there were three passenger services in each direction during the summer and two in the winter, along with the return of The Ghost to Glasgow. There were no timetabled passenger services on a Sunday, although during the summer months there were Sunday Specials and regular excursions to deal with.

The North British summer timetable for 1922 shows down trains from Glasgow stopping at Rannoch Station at 8.57am, 2.48pm and 7pm. Gorton and Corrour were not on the timetable at that point, although these services called at Corrour 16 minutes after they left Rannoch.

The first sleeper services to and from London Kings Cross from Fort William were introduced during the summer months only on July 22, 1901, and in October 1929 sleeper cars began operating throughout the year.

The LNER's last West Highland Railway summer timetable, published in 1947, shows just two passenger services from Glasgow stopping at Rannoch, the sleeper at 8.58am and the last train of the day at 7.02pm. Both paused briefly at Corrour at 9.14am and 7.18pm respectively. A Saturday only service leaving Glasgow at 11.25am and terminating at Fort William at 3.49pm was not scheduled to stop at either.

In the opposite direction, the 7.46am service from Mallaig stopped at Corrour at 10.43 and Rannoch at 10.57am while the 1pm sleeper train called at Corrour at 4.04pm and Rannoch at 4.18pm. A Saturday only evening train from Fort William to Glasgow was not timetabled to stop at either.

British Railways winter timetables for 1948 show down trains calling at Rannoch at 8.58 (the sleeper from London Kings Cross) and 6.57pm, then Corrour at 9.14am and 7.13pm respectively.

Travelling in the opposite direction, the Glasgow service from Mallaig via Fort William called at Corrour at 10.33am and 4.03pm and Rannoch at 10.47am and 4.17pm. The latter had sleeper cars destined for London.

In addition to the sleeper service, travellers could later take their cars on the train to Fort William thanks to British Rail's Motorail service. Launched in June 1955 between London and Perth as *Car-Sleeper Limited*, it was re-branded in 1966 and introduced to the West Highland Railway on May 14, 1990. The service was, however, short-lived. While sleeper trains from London to Fort William survived, Motorail was withdrawn just five years later.

When the signalmen at Corrour, Rannoch and Gorton were not attending to passing trains or dealing with the other business of the day, they were kept busy cleaning and maintaining their stations, digging out weeds, raking the platform gravel and undertaking minor repairs. The NBR made regular unannounced inspections to keep them on their toes.

6

Life on the Moor

Victorian and then Edwardian travellers drawn to the West Highland Railway by the NBR's colourful posters extolling the majesty of the scenery must surely have been captivated by the views from the carriage windows as their trains rattled north, past Loch Lomond and then on over Rannoch Moor towards Ben Nevis and Fort William.

The lavish pictorial promotions promised a journey 'among the mountains, lochs and glens in the West Highlands of Scotland' with the route 'embracing the largest portion of the finest scenery in Scotland'. Quite a boast and little wonder that excursions and Sunday specials proved so popular.

In the early years of the line, the Craigendoran Route and West Highland Route were two such outings. The first, described as an 'expeditious and picturesque' trip, offered connections with steamers sailing to the islands of Iona and Arran while the West Highland Route linked with vessels to the Outer Hebrides and Inverness via the Caledonian Canal.

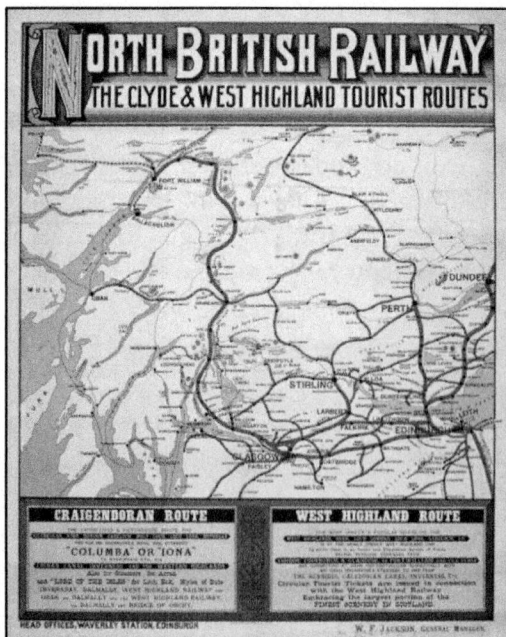

Journalists were equally enthusiastic. Ahead of the opening of the line, the *Dundee Courier* published on April 14, 1894, a lengthy review, the writer stating: 'It is no exaggeration to say that once the West Highland Railway is constituted an integral part of the great North British system, a new field will have been unfolded for the lover of Scottish scenery to revel in.'

A North British Railway poster extolling the scenic virtues of the West Highland Railway

'The railway not only covers large tracts of country untouched by the recognised coach routes, but it provides facilities for an endless number of new circular tours through some of the most romantic regions of the country,' he added.

The writer had enjoyed what would now be described as a Press trip over the line at the invitation of the NBR and engineers Forman & McCall. His party travelled in an observation car linked to three engines, the journey organised ostensibly to test the line's viaducts.

Passengers included London-based artist Edward Slocombe who was employed to pen sketches of scenic spots for the railway company's official guide, *Mountain, Moor and Loch*. Slocombe spent several weeks travelling up and down the line on contractor's trains.

Described as a 'genial' soul by the *Dundee Courier* correspondent, he confided that 'during his short sojourn in these parts he had found great difficulty in keeping himself supplied with tobacco'.

LNER D34 4-4-0 Glen Class locos 9407 *Glen Beasdale* and 9307 *Glen Nevis* approach Rannoch Viaduct on August 3, 1930 (H C Casserley)

The newspaper previewed what passengers crossing Rannoch Moor could expect, revealing: 'The vista presented to the traveller becomes wilder and wilder until he emerges on the desolation of the Moor of Rannoch. For a distance of some 20 miles, he is carried across the silent waste, bounded on either side by the awe-inspiring grandeur of the mountains, with a magnificent view of Schiehallion on the east and, every now and then, a glimpse of the snow-capped Ben Nevis on the west.'

Other newspaper reporters were equally positive and there were glowing write-ups of both the line and its formal opening ceremony and, after clearing the hurdles thrown up by the first winter of operation, tourists did indeed flock to travel its tracks. Crossing Rannoch Moor, one wonders exactly what these Victorian trippers drawn north from Britain's towns and cities made of the wilderness.

In August 1900, a correspondent from the *London Daily News* joined one of the popular circular tours of the time, taking the steamer from Glasgow to Oban and then on up Loch Linnhe to Fort William, boarding the West Highland Railway for the return south where he was surprised to discover stations not marked on the railway map or included in the timetable.

Trundling past Loch Treig, he wrote: 'We are still climbing when suddenly on go the brakes as a huge noticeboard bearing the legend "Summit of Line, 1350ft" presents itself and a few seconds later the train is at rest at a baby island platform.

'What station is this? None is shown in map or timetable. As a matter of fact, it is Corrour crossing station, to which passengers are not booked, although they may join or alight from the train while the electric tablets are being exchanged, paying the difference to or from the station on either side.'

With typically Victorian gusto, the journalist alighted at Corrour and, to experience Rannoch Moor for himself, set off on foot, walking the permanent way to Rannoch Station.

'It is difficult to realise solitariness whilst seated in a luxurious corridor carriage but tie-counting on this desolate stretch of track impresses the pedestrian forcibly with the true meaning of the word. With the exception of a gang of surfacemen gauging the curves, not a living soul is to be seen the whole way,' he added.

Doubtless others were equally curious. Passing through Gorton, Rannoch and Corrour, perhaps spotting the signalman or stationmaster on the platform, they must have pondered how people existed in such barren isolation, in a land where roads were rare, houses were few and far, and where there was no evidence of any of the amenities of the urban life they knew.

They need not have worried for the railway company had it all in hand, providing everything its staff and their families needed to survive here, from deliveries of fresh milk to schools for the children.

The 1870 Education Act ruled that every child in Britain should receive elementary schooling between the ages of five and 12 and the railway children were no exception. The provision of education for the offspring of the NBR stationmasters, signalmen and surfacemen of Rannoch Moor varied over the years, depending on demand.

At Corrour, the children of estate workers were educated at a small school established adjacent to the lodge at the head of Loch Ossian. The estate employed a teacher, who lived there. The signalman's family, however, were not educated there and his children attended classes in either Fort William or at a small school in Tulloch.

Donnie MacNicol was born at Corrour Station on May 23, 1923. The youngest of eight – he had six brothers and a sister – and he recalls attending school in Fort William.

'The train arrived at 9am. We could see and hear it coming from Rannoch. Other children came from Rannoch and Gorton. There was a big family at Gorton,' he said.

'There were folk from Tulloch, Roy Bridge and Spean Bridge who joined the train. It was a big, big crowd and of course we got into trouble at times running up and down the corridors.'

The children took sandwiches to sustain themselves on the journey, arriving at Fort William Station at around 10am.

'We all trooped into our classes an hour late,' Donnie continued. 'In the junior school they were very good in that they kept us up to date. There was religious instruction for the first hour in the morning which we missed so it didn't affect our education.

'But, when we went into higher grade, we didn't get that and we missed a lot. Imagine missing five hours a week, all through the school year. It is quite a bit we missed so when the exams came some questions we just didn't know.

'School came out at 3.55pm and the train left at four, so we didn't dilly dally. We had to run down and catch it. There was an odd occasion when somebody missed it, but friends took them in.

'Then it was home and by the time we did our homework and had a meal that was the day. You always had plenty of homework to do. It was handy in the morning on the train going down if you were stuck with something you could catch up with your homework,' Donnie added.

Duncan Mackinnon, who lived with his family at Luibnaclach for 20 years from 1948, recalls spending many hours in the signal box at Corrour waiting for the train to take him to school in Tulloch.

Marion Barrie, daughter of a gamekeeper, attended the school at Corrour Lodge up until its closure in 1943 after which she moved to the school at Tulloch. However, rather than travel daily by train with the railway children, she made the journey on a Monday and boarded with the teacher, returning home at the end of the week.

Rannoch Station School (Author)

Up the line at Rannoch Station, a new school was built for the children. To enable quick and cheap construction, a corrugated iron kit building was purchased, brought north on the railway and erected on a plot of land 250 yards east of the station, by the B846 road. It included a classroom and adjacent house for the schoolmistress and could accommodate up to 30 pupils.

Opened around 1900, the school initially took children from Gorton who travelled to and from their lessons by train each day. However, in 1938, with the school running over-capacity and Gorton located in Argyllshire, the Perthshire Education Authority handed responsibility for the pupils back to its neighbouring county and the education committee of Argyll was forced to come up with an alternative.

They contacted the LNER, asking if they could erect a temporary school either in the siding at Gorton or on the platform, using one of the company's carriages. An old Great North of Scotland Railway coach body was duly sourced, fitted with desks and chairs, and positioned on the platform.

Safety barriers were installed by the doors to prevent energetic youngsters running straight out and falling on to the track and a toilet block for the pupils was erected at the southern end of the platform. In March 1938, with an intake of eight young scholars, the new school opened, its teacher travelling daily by train from lodgings in Bridge of Orchy. The school appears to have remained in operation until the late 1950s or early 1960s.

Teacher and pupils on the doorstep of the carriage school at Gorton

Rannoch Station School closed in 1959 and thereafter primary school pupils travelled north once again to Tulloch, which appeared to be more convenient than making the road journey east to schools at Bridge of Gaur or Kinloch Rannoch. The corrugated iron building remains and is now a private dwelling.

With time spent travelling to and from school and homework to complete, after school activities were limited. During the week, bed and an early start beckoned. However, at weekends, the children were free to enjoy themselves.

Donnie MacNicol recalls that on a Saturday he and his brothers and sister would go out on their bikes, cycling along the Loch Ossian track to buy lemonade, cola, American cream soda and Lyon's cupcakes from the small shop at Corrour Lodge.

The carriage school on the platform at Gorton, looking south along the platform towards the toilet block and signals (Colour-Rail)

'That was our delights. It was great,' he said. 'We did a lot of cycling and fishing – there were plenty of trout in the lochans – and we played shinty on the platform. We had no shinty sticks so we had to cut a bit off the tree.

'We played football too. The shed had a window and it got broken so often because we played football against it my father had to board it up.'

The railway also took care of the practicalities of everyday life. Fresh milk was delivered in churns to Gorton, Rannoch and Corrour on the first down train of the day while at Gorton drinking water also had to be delivered as the brackish moorland water was deemed unfit for human consumption.

The signalmen and stationmaster's wives were issued with passes to travel once a week to Fort William to buy provisions.

Donnie MacNicol recalls: 'My mother went shopping to the town every Friday and a big box came back with the messages. Of course, we had no fridge, but we managed.'

There were occasions, he remembers when food ran low, but the families supplemented their store cupboard purchases with vegetables grown in their gardens, eggs from their own hens and, at Gorton, milk from goats. They also fished, caught rabbits and, from time to time, chanced upon game.

'Once we had nothing for the evening meal and when I went home two grouse had hit the wires and were lying dead. We never wanted for anything.'

Corrour stationmaster Archibald MacNicol and LNER J37 0-6-0 loco 9454 on a southbound freight service (Donnie MacNicol)

Children from both Corrour Estate and the railway who stayed on at school beyond the age of 12 boarded during the week in Fort William, travelling down on Monday morning and returning home on Friday evening.

Special occasions were few and far between at the railway schools but there were trips out from time to time and, in December, Santa Claus, travelling aboard a special train, visited the pupils at Gorton and Rannoch, delivering presents.

Donnie's father, Archibald, was stationmaster from 1908 to 1941 and during that time there were various changes at Corrour. In the year he started, his employer, the NBR, which had until that point operated and, to a point, subsidised the West Highland Railway, took the whole operation over, branches and all.

Closer to home, in 1921, a telephone line was installed between Corrour Lodge and the station and the stationmaster was paid £5 a year to take and receive the estate's calls, a particularly busy job during the shooting season.

Communications were further improved with the construction of a road running along the side of Loch Ossian from the boatshed, which was converted into a youth hostel, and the lodge.

Motor cars and vans were delivered by rail and these were used to convey people, letters, parcels and goods from the station to the lodge. In the 1920s and 30s the fleet included a Morris Oxford saloon, a three-ton Fordson eight-cylinder lorry, a small Austin Seven and an Austin Twelve shooting brake.

Regular consignments carried over the lochside road included supplies for the shop run by the estate and books for a tiny public library maintained by the housekeeper.

At any one time there were just 10 titles to choose from but, thanks to the train, these were constantly changed.

Road along south side of Loch Ossian (Author)

Donnie continued: 'My father had to get up at half past four in the morning because a goods train called The Ghost came through early. He was there all day until 7pm. There were perhaps three passenger trains each way and lots of fish trains and goods trains so he was kept going all the time.

'He had to go up very steep steps to that signal box but it was a lovely station. It had to be because there was an inspector who came round and if he saw as much as a matchstick on the platform there would be a row about it.

'There was an outside tap at the end of the building and the brass tap had to be polished regularly and the platform had to be kept tidy. I remember my father raking the gravel regularly,' he added.

The MacNicol family would have seen the return of navvies to the West Highland Railway, drawn this time by the Blackwater Dam construction scheme.

The last major construction project in Scotland to be built without the aid of machinery, the mighty dam, eight miles to the west of Corrour Station as the crow flies, was built between 1905 and 1909 by up to 3000 men, again mostly from Ireland. It funnelled water down the glen to a power station at Kinlochleven Aluminium Smelter, the North British Aluminium Company's second plant in Scotland (their first was at Foyers, on Loch Ness).

With no surfaced roads to Kinlochleven, many of the navvies tramped mountain tracks and trails to the work camps located at the head of Loch Leven. Rough roads led in from Glen Coe, to the south, and Fort William, to the north.

The savvy navvy, however, took the train north from Glasgow to Corrour and walked to Lochtreighead from where a good path led down through the glen to the dam site. As a result, Corrour quickly became popular with the navvies and, despite its private halt status, many alighted at the station.

Those tempted off the trains one stop earlier at Rannoch, perhaps by the hotel bar or shop with its promise of wines and spirits, faced a less certain fate. The hike via Loch Laidon was longer, while a more direct crossing following the Black Water burn was pathless and the terrain very rough underfoot. Some, sadly, did not make it, getting caught in storms or succumbing to exposure and hypothermia on the moor as they attempted their traverse.

On July 15, 1907, newspapers reported the discovery to the north-west of Rannoch Station, by the county boundary bordering Lochan a'Chlaidheimh, of two badly decomposed bodies. The first was found on the Friday lying in the lea of a boulder by a workman who was walking across the moor to Kinlochleven.

'Making his way to the railway, he trudged to Rannoch and there reported the matter,' the *Port Glasgow Express* reported. 'Mist and darkness settling in, however, the search had to be postponed until Saturday.

'The informant accompanied the search party, but for a time had a difficulty in locating the spot where the body lay. In the meantime, Inspector Chisholm, Fort William, was somewhat taken aback by coming across the skeleton of a second man.'

In addition to the two bodies, which lay 500 yards apart and were little more than bleached bone, torn flesh and tattered clothing, officers found a purse containing 10 shillings and a small bundle of personal effects.

'The men had presumably been making for or returning to the Kinlochleven works and had been overtaken by severe weather or lost their way in the vast undulations of this isolated moor,' the newspaper concluded.

The remains were buried where they lay.

The construction of a third aluminium smelter in Fort William, completed in 1929, saw navvies cross the moor once again, but this time they could remain safely on the train all the way to the northern terminus without need to step out and brave the wild moor on foot.

Passenger numbers leapt from 23,400 in 1923 to 31,000 in 1926 and 43,250 in 1928 while the signalmen and women at Gorton, Rannoch and Corrour experienced a significant upturn in freight, from around 10,000 tons a year before work on the smelter and associated hydro-electric development started to over 47,000 tons a year at the peak of the project. Once complete, there were regular trains carrying alumina from Burntisland, in Fife, and later Invergordon, in Ross & Cromarty, to the smelter. These services continue to this day, although the alumina now comes from a terminal at North Blyth, Northumberland.

While Corrour enjoyed an upsurge in passengers during the Blackwater Dam project, Rannoch Station benefitted from a healthy increase in business during the construction of hydro-electric power schemes in Perthshire during the late 1920s, 1930s and 1940s. The Tummel Bridge and Rannoch project involved construction of a dam and power station on the River Gaur, a dam at the southern end of Loch Ericht and a power station on the northern shoreline of Loch Rannoch.

Northbound ex-NBR 4-4-0 62496 *Glen Loy* at Rannoch Station on September 9, 1959, approaching the signalwoman to exchange tokens. Note the Renton Stone behind her (Colour-Rail)

Once again, there was a large influx of men. In May, 1928, the *Dundee Courier* reported that several hundred were at work on the Loch Ericht 'electric works', their food and supplies being delivered to Rannoch Station, 12 miles to the west of the construction site and camps, with many more workers expected to follow.

The influx of navvies was not without its problems. The public bar at Rannoch Station proved to be a popular watering hole with labourers and newspapers reported on a variety of sheriff court appearances centred on offences committed at the pub. These ranged from assaults to thefts and, invariably, the accused gave their addresses as 'The Huts, Rannoch'.

After a sustained period of heavy industrial use, the West Highland Railway, which was absorbed by the LNER on January 1, 1923, saw leisure traffic grow in the 1930s as more and more people from Scotland's towns and cities discovered the delights of the great outdoors.

Fishing had long been popular on the lochs and rivers of Rannoch and Tummel. The moorland lochs offered plentiful catches of brown trout while, to the east, trout, charr and coarse fish like pike were targeted. Those hoping to reel in a salmon looked to the Tay and Tummel rivers.

The arrival of the railway meant that, for the first time, anglers from Glasgow could travel up in the morning, enjoy a full day's fishing, and then return home on the last train.

As a result, Rannoch Station was one of the busiest halts on the line. Figures for 1930, as an example, show the station was second only to Fort William in terms of passenger numbers with over 11,000 people passing over its platform, earning the LNER revenue of almost £4000 a year.

Rannoch Station Hotel (later the Moor of Rannoch Hotel) offered the visiting fishermen bed, board and permits for Loch Laidon and, from the opening of the line for many years thereafter, there was a daily motor coach service from the station to the Dunalastair Hotel, in Kinloch Rannoch, at the eastern end of Loch Rannoch.

During the summer months, Fisher's Hotel, in Pitlochry, ran a daily coach via Loch Tummel and Loch Rannoch to Rannoch Station, leaving the hotel at 9am and departing from the station for the return journey at noon. From the late 1960s until 2011, the Royal Mail operated a Postbus over the same route.

Rannoch Station Hotel (later the Moor of Rannoch Hotel)

These services enabled travellers who had come north on the West Highland Railway to return south via the Highland Railway, boarding the train at Pitlochry Station.

During the stag stalking and grouse shooting seasons, Gorton, Rannoch and Corrour stations were all busy. Between August 12 and October 12, when grouse are shot, all three employed additional clerks – young women paid 12 shillings a week – to handle the traffic. This included receiving the shooters and dispatching their game to the markets of the south. At Rannoch Station, a chute was erected to load deer carcasses into waiting vans.

With restaurant cars added to the West Highland fleet in the summer of 1929, replacing the refreshment baskets that had traditionally been available at Arrochar & Tarbet and Crianlarich, Sunday excursions took off with packed trains to Fort William and Mallaig dispatched north from Glasgow, Edinburgh and other points on the LNER network.

In 1933, the company was the first to introduce static, self-catering camping coaches, old passenger carriages no longer deemed fit for active service which were converted and deployed to otherwise disused sidings.

In 1934, as part of a second distribution of 25 coaches, one was sent to Rannoch Station where it was parked to the east of the platform, by the site of the former turntable. There were no mains services and campers had to visit the station to use the toilets and collect water. Paraffin was provided for cooking and lighting. Despite these limitations it was enormously popular but the coach was withdrawn in 1939, at the start of the Second World War.

While it was never reinstated, another camping coach was deployed to the siding, this one remaining there to this day. It was not intended for holidaymakers but for use by members of the angling section of British Rail's Staff Association.

A former dormitory coach was acquired and converted for its new role. Equipped with 12 bunks, heating, lighting and cooking facilities, lockers, tables and chairs, it arrived on site around 1954. Over the years the external appearance of the carriage has altered and latterly it has been reclad and painted green.

Frequented over the years by railwaymen from different British Rail depots, it has been referred to as the Polmadie camping coach, the Cowlairs anglers' coach and the St Rollox Works camping coach.

Ewan McRae recalls regular weekend trips to the coach with his grandfather during the mid-1960s. He said: 'At the weekends and during the school holidays my grandpa would often take me to the Polmadie railway camping coach where on each occasion we would spend a few unforgettable days and nights. We'd catch the train from Queen Street to Rannoch Station and wander over to the old coach.

'It was a grassy spot, surrounding by silver birch trees and there were old caravans dotted about. The coach itself was next to an old turntable. It was long gone but you could still see the circle of stone where it had been.

'I remember the night fishermen would come back and awaken us with their freshly caught trout from Loch Laidon. They'd be gutted and fried in butter, accompanied by eggs, bacon and black pudding, all washed down with strong, sweet tea, topped off with evaporated milk.'

Further north the youth hostel created from the former boatshed at Loch Ossian in 1931 offered equally basic, no frills accommodation to early walkers and climbers drawn to the mountains bounding Rannoch Moor. In August 1934, writer Ian MacPherson extolled its virtues in the *Aberdeen Press and Journal*.

'The chief and shining merit of Loch Ossian is that it is not merely a halting place, providing shelter for a night, but lies in the very heart of a path-seamed region of the most varied and magnificent kind where one can issue daily for weeks in different directions without exhausting either the varieties of path or the pleasures which each separate route provides,' he wrote.

'But the moor is always dangerous, not from its hags and pitfalls alone. Sudden changes of weather, wisps of mist that confuse the clearest head and obliterate all idea of direction, make discretion chief virtue.

'Yet this empty scene has its fascinations. Everywhere there is ample space and room enough, no crowds elbow one, wild birds cry, the hags and gullies recall the world's original state, as if man had never existed.

'Although one comes here for only a few days, one feels something of that which makes the people of the country prefer it above all other regions. Rannoch Moor possesses a strange character, a wild insularity, a freedom and emptiness which no other district has. Until the hostel was provided by Loch Ossian one could scarcely risk traverse of the range and moor. It is made safe, and as easy as a thing worth doing should ever be.'

On September 15, 1934, perhaps in response to the growing popularity of Corrour as a destination for outdoor pursuits, the private halt finally took its place on LNER timetables as a fully-fledged public station. In practical terms, nothing changed but at least now passengers could buy a ticket for Corrour – rather than for Tulloch or Rannoch, depending on which direction they were travelling – and see what time their trains were due to either arrive or depart.

Gorton did not benefit from such openness and remained out of bounds. Over the years, journalists, their curiosity piqued by this strange anomaly, attempted to visit but while their trains came to a halt they were routinely rebuffed by staff.

The Second World War ended Sunday excursions, the motor coaches to Kinloch Rannoch and Pitlochry and the heady days of the camping coach but the West Highland Railway soldiered on and, in time, the tourists would return.

Stationmaster Archibald MacNicol (centre) with a driver and fireman on the platform at Corrour (Donnie MacNicol)

7

War & Winters

During the Second World War there was an increase in military traffic over the West Highland Railway. While much of it was confined to the southern portion, the trains servicing a new naval base at Faslane on the Gareloch, a second naval installation was established at Corpach, west of Fort William, and there were significant movements of timber south, much of it coming from the Fort Augustus branch.

In April 1942, Corrour Lodge was destroyed by fire. The blaze was not started by a stray enemy bomb, but by a blowtorch. A plumber repairing burst pipes accidentally ignited insulation in the wall and the flames quickly spread through the pine panelling. With no fire brigade close at hand, very few estate workers on the ground and untested firefighting equipment installed, those who did try to battle the blaze found the water pressure was too low. The house burned to the ground and most of its contents were destroyed. Fortunately, no one was killed or injured.

Ironically, Sir John was in the process of moving some of his more valuable paintings north from his stately city home, Pollok House, in Glasgow, to the lodge for safekeeping. By a lucky twist of fate, they had only made it as far as Corrour Station and were still safely packed into railway vans parked in the siding when the fire broke out.

Sir John discussed a replacement with architect Reginald Fairlie, who had altered and extended the lodge in 1935, and although plans were drawn up for a truncated version of the original, post-war austerity and high taxes ensured this was never built. After the war, during visits to the estate, Sir John stayed in the head keeper's house, until his death in 1956.

What remained of the fire-ravaged structure was demolished in the 1950s and was replaced by a wooden three-bedroom bungalow in 1958.

For 19 years from 1944, Sandy Thompson was stationmaster at Corrour, living with his family in the trackside cottage. Sandy's experiences were documented in a short BBC film, *The Station on the Moor*, which was recorded and broadcast in 1965.

Born in Joppa, on the outskirts of Edinburgh, and brought up in nearby Millerhill, Sandy left school at the age of 14 and started on the railways as a train register boy at Millerhill signal box. After a short period at Berwick, he was put in charge of his first signal box, Esk Junction, near Edinburgh.

In 1925, he moved to Mallaig Junction, before transferring to Arrochar & Tarbert and then Crianlarich Upper where he remained until 1944 when the post at Corrour came up.

'I was very keen to get a stationmaster's job and had thought that if I got the position at Corrour then I would have the necessary qualifications to allow me to get a stationmaster's job,' he told the BBC.

'When we got to Corrour, we got to like the place and after that I forgot all about stationmaster's positions. We just wished to remain in Corrour.'

Sandy, his wife Pollie and their three children arrived in the depths of winter – there were three feet of snow on the ground to welcome them – but soon settled in. His children attended Tulloch School, leaving home every weekday morning on the 9.14am train and returning at 4pm, while his wife was employed as signalwoman.

The station house at Corrour under snow

The couple worked long days, starting at 3am and continuing until the final train of the night cleared Tulloch at 7.50pm. Before breakfast they dealt with three goods trains and a passenger service, accepting each from Rannoch Station before handing it on to Tulloch.

At 9am, the Post Office opened with mail – which arrived on the 9.14am train – to sort, postal orders to sell and telegrams to send. In those days, a lot of Corrour Estate's shopping was done by post and the lodge postman would visit each morning.

The stationmaster also oversaw the loading of timber on the siding and the unloading of estate supplies, all of which arrived by rail, along with a steady stream of passengers, the majority of whom came during the shooting season.

In common with his predecessors, Sandy's most arduous shifts were during the winter months. Steps may have been taken to better protect the line from drifting snow following its first chaotic winter of operation, but the season remained a particularly testing time, the exposed moor and its stations frequently battered by the elements.

The winters of 1902, 1906, 1908 and 1909 all saw engines or trains brought to a standstill by snow on the moor, the crews and their passengers often blocked for several hours. In January 1909, passengers in one stormbound train spent two days and two nights trapped in their carriages at Rannoch and then Corrour.

The *Dundee Courier* of January 18, 1909, reported on the severe snowstorm, one which its Fort William correspondent had not seen the like of 'for many a long year'.

With snow lying up to three feet deep, the Friday evening service from Fort William to Glasgow managed to get through, albeit several hours behind schedule, after being held up at Gorton.

However, as the wind picked up, a 'brilliant display of lightning and stinging showers of hail' heralded the arrival of fierce blizzards and the northbound was less fortunate.

It left Crianlarich behind a snowplough and, after battling through the snow, made it past Gorton to Rannoch Station. Here, however, with the cuttings ahead reportedly filled with snow – up to depths of 20 feet in places – it was stopped and held for the night.

'There were about 20 people aboard and the prospect of spending a night in the heart of the trackless Moor of Rannoch, the largest and most desolate in Scotland, can be better imagined than described,' the newspaper added.

'As the carriages of the West Highland system are heated by means of steam from the engine, the travellers suffered no discomfort in this respect, but the pangs of hunger were not so easily assuaged.

'The railway officials did what they could in the way of supplying hot tea and what eatables could be spared, but naturally they could not be expected to have ample supply for such number.'

Spirits remained upbeat. One passenger shared out a cake and bottle of wine he had with him and played his bagpipes at intervals through the night. Come morning, the train steamed north once again. It made it through the snowshed but ground to a halt as it neared Corrour.

'Here the hardships and experiences encountered at Rannoch were enacted anew, but the spirits of those on the train were less buoyant,' the *Dundee Courier* observed. 'The want of sleep and absence of proper food had a depressing effect and this, coupled with the lonely and isolated location, made existence for the time extreme.

'The solitary signalman did his best for the done-up folk aboard and some more daring than others annexed a case of sausages from the van and had these prepared as delicacy.'

In other words, some of the fed-up, hungry passengers pinched a box of sausages, and who could blame them as they prepared to spend their second night on the moor, this time without the luxury of either wine or cake.

The train remained stuck at Corrour until 11am on Sunday when ploughs and large squads of men working from the Fort William end managed to finally clear the track. Shortly before 1pm it eventually reached its destination.

While the travellers on this train helped themselves to some sausages without consequence, others were less fortunate. On another occasion, driver Jock Campbell and his fireman, a character known as Wee Barry, found themselves snowed in at Rannoch Station. Stranded for two days, Barry decided to dip into the contents of the 'road van', a van attached to the passenger train carrying supplies for hotels and shops in Fort William.

He released a case of whisky and some biscuits and shared them amongst the passengers. Unfortunately, word of this reached the LNER and Barry was arrested. When he appeared in court to answer the charges, the case

The snowstorms of February 1951 brought a relief train from Glasgow loaded with food and railway gangers to a halt on Rannoch Moor, just a mile and a half from the passenger service it was sent to rescue. So deep were the drifts that it had to be hauled back by a second relief train (London Illustrated News)

was swiftly dismissed, the sheriff remarking that, in the circumstances, the company was lucky that a few bottles of whisky was all it lost.

Donnie MacNicol, whose brother Johnny became station master at Spean Bridge, remembers that during the worst winters of the Second World War, 'trains were delayed very, very much.'

He said an old steam engine filled with concrete and pushed by a steam engine, with wagons of concrete behind, formed an effective snowplough.

'During long snowstorms, two engines and two ploughs were linked together. With an engine and a plough at both front and rear, it cleared snow in either direction. Once the line was cleared, it was just a case of keeping it clear by running back and forth,' he added.

Gorton signal box and the carriage school under heavy snow

Sandy Thompson said life on the line was pretty grim during the winter, but they just had to get on with it.

'I've seen five or six surfacemen and sometimes the flying squad out there to keep the points clear and let the trains through,' he continued.

Carriages stranded on Rannoch Moor

'It was impossible for us really to work shifts about because there were times it was impossible for any person, especially a woman, to be working in the cabin during snow and blizzards.

'It was very difficult to see anything in the snowstorms. It was just blowing like smoke and it was very choking.

'The winter of 1946/47 was a very bad one, the deer lying dead in their groups here and there down the back of the snow fences. It was just blowing a hurricane.

'I worked 24 hours a day for almost three weeks in 1947. But the trains got through. The snow ploughs worked all night, just one after another running backward and forward between Crianlarich and Fort William.'

His son Alasdair Thompson added: 'You come around a corner and it goes into a small cutting and the snow might be eight or ten feet high but two hundred yards long. This merely slows the plough down. But near Rannoch station the cutting passes through rock.

'The snow didn't get thrown aside because there were rocks to right and left, and it piled up at the front. This is how a lot of trains got stuck.'

During Christmas 1947, a passenger train was snowed in for two days at Corrour and Alasdair remembered how his parents offered shelter to two passengers, a woman and her baby.

'They took them into the house and gave them my wee sister's room. We just shared all the provisions together. Mum shared them with the dining car because they were running short.'

Four years later, in February 1951, 23 passengers spent an unplanned night stranded at Corrour after their train, the 3.46pm Saturday afternoon service from Glasgow to Fort William, ran into a 10-foot deep snow drift.

A squad of 70 men dug through a 100-foot-long bank of snow but it was 30 hours before the train was freed by a relief engine and hauled to Fort William. Earlier in the day, a relief engine had also become stranded.

Arriving in Fort William one of the passengers, local businessman Murdo McKenzie, told waiting reporters: 'It was the most terrifying snowstorm I have ever seen. A howling gale swept the snow into huge drifts, some of them 15 feet high. We daren't open a window to look out. One man tried it and his hat whipped off his head and was blown across the line on to the moor.

'The first night was the worst. The engine's fire had to be put out when the water supply failed. The dining car staff did their very best for us though and gave us tea and hot soup. We came out of our compartments and stayed in the dining car, where it was warmer with the heat from the kitchen.'

Snow was not the only problem on the West Highland Railway. It has also been brought to a standstill by heavy rain and landslips. Alasdair Thompson recalled a day when the snowplough from Fort William was delayed. Travelling south by Loch Treig, it had struck a seven-ton boulder, derailing as it rode across the stone and tipped on to its side. Fortunately, no one died, and the crew managed to walk up the line to Corrour, where his mother Pollie tended to the fireman's burns.

In an accident on the night of July 6, 1894, fireman James Tucker was less fortunate. During a heavy rainstorm, he was working on the footplate of a locomotive hauling 15 empty wagons south. The train successfully traversed Rannoch Moor and, passing Gorton, trundled on through the downpour towards Bridge of Orchy.

Rounding a bend, Tucker and the driver spotted a gap in the embankment, rails and sleepers hanging in the air. The permanent way had been washed out by floodwater tumbling off the hillside. There was no time to stop. The locomotive left what remained of the track, falling into the void, the tender tumbling down on top of it. While the driver escaped the accident with a sprained ankle and bruised arm, Tucker died at the scene.

In more recent times rockfalls on the steep slopes above Loch Treig have claimed trains but luckily not lives. On June 28, 2012, a GB Railfreight Class 66 locomotive, 66734 *The Eco Express*, hauling 21 wagons of alumina powder and three empties from North Blyth to Fort William hit a boulder and derailed. The locomotive and first five wagons left the track, the engine sliding down the slope towards the loch. Badly shaken, the driver was airlifted to hospital but escaped serious injury while the engine came to rest against a shelf just short of the shoreline.

Engineers were able to re-rail and remove the wagons, which were initially shunted up the line on to the siding at Corrour Station, repair the track and re-open the railway within a week, but they were unable to resurrect the locomotive and it sat partway down the embankment, covered and secured by cables, for several months while GB Railfreight tried to work out how best to rescue it. Unable to find a cost-effective solution to retrieve the £2 million engine in one piece, it was written off and dismantled on site for spare parts.

Back in the 1940s, Sandy Thompson's memories turn to more domestic matters. The family's groceries arrived by train and there were occasional shopping trips to Fort William. If anyone in the area fell ill, word would be sent to Tulloch and the railway company would send an engine and coach with a doctor or a nurse.

Sunday was the only day off and with no church nearby the family would listen to a service on the radio before venturing outdoors.

Sandy said: 'We would be hiking in the hills or walking round the loch. We would go to the top of Beinn a'Bhric where there was a big rock called the Witch's Chair which was shaped something similar to a chair and it was the custom to sit there and wish a wish.'

While Sandy also enjoyed shooting and fishing, his children made their own amusements, spotting wildlife, sledging on iced pools during the winter and fishing in the summer.

'We did feel it a bit lonely at times, but we just knew we couldn't get away and we had the wireless which kept us,' he added.

As time passed, Sandy witnessed changes on the line, most notably the replacement of steam locomotives with diesel engines.

'We missed the old steam engines. I think it was more homely seeing the old steam engine coming chugging round the corner,' he recalled.

Eventually the time came to bid Corrour farewell and in 1963, with their children grown up, Sandy transferred to the signal box at Corpach and he and his wife moved into a house in the nearby village of Caol, something of a culture shock for the couple.

'I did like Corrour,' he said. 'We find it's a big change from being in the moors by ourselves. We were never in the habit of having so many neighbours round us. We took it strange at first.'

New incumbents took their place and while life at Corrour Station continued pretty much as normal, major changes occurred on the neighbouring estate. In 1966 three-quarters of the land was sold by the Maxwell family to the Forestry Commission who built a private road from Torgulbin, on the A86, to Corrour Lodge, laid a track along the north side of Loch Ossian and repaired the old road along the south side of the loch. In 1972, for the first time, Corrour Lodge and Corrour Station could be reached by motor vehicle.

In the latter years of the decade, the 80-year-old staff cottage at Corrour Station was declared no longer fit for human habitation and demolished. It was replaced by a prefabricated two-storey structure, a sign perhaps that British Rail had no long-term plans to maintain a human presence at the remote outpost. A similar building was erected up the line at Glen Douglas, again to replace the original condemned station house.

The Maxwell family bought Corrour Estate back in 1984 but were unable to meet the costs of its upkeep and, in November 1995, it was put up for sale in its entirety.

Motive power may have moved on from steam to diesel but, as this image from the early 1980s shows, the weather on Rannoch Moor could be just as severe.

8
Maintenance

To keep passenger and freight trains running safely and efficiently over Rannoch Moor, the NBR posted surfacemen (the Scottish name for platelayers) at regular intervals along the route. Each was responsible for their own stretch of track, from going out before the first passenger train of the day to check that no obstructions had fallen on to the line during the night to checking the sleepers, rails, chairs and ties were in place and ensuring the ballast was evenly spread.

The early Rannoch Moor surfacemen were provided with lineside cottages where they lived with their families or, in the case of single men, lodged. There was one at milepost 52½, near Achallader to the south of Gorton, one at milepost 58, to the north of Gorton, two at Rannoch Station and two at the remote outpost of Lubnaclach. Huts, commonly referred to as bothies, were erected at intervals along the track to provide shelter and these were fitted with stoves or fireplaces.

Working on the track was – and still is – a potentially dangerous job and, in common with lines around the country, there were fatalities on the West Highland Railway. In November 1895, surfaceman Donald Campbell was struck and killed by a train that approached from behind as he adjusted rail joints near Bridge of Orchy.

In December 1899, track worker James Scanlan, a young single man, was run over and killed by a train near Ardlui. The locomotive struck his foreman a glancing blow, but he escaped injury.

Early on the morning of January 1, 1910, Rannoch-based foreman surfaceman Alexander Campbell and surfaceman Thomas Campbell were knocked down and killed just south of Rannoch Station by a special train en route from Fort William to Crianlarich.

The subsequent inquiry heard that it was so dark at the time that the driver could not see beyond the front of his engine. The train jerked, he applied the brake and, on inspecting the track behind them, the guard found the two bodies, bloodied and covered in mud. The jury apportioned no blame.

In equally poor visibility, this time the result of a snowstorm, Rannoch Station-based ganger William Grant was knocked down and killed by a train while he worked on the track near the Gaur Viaduct on December 16, 1936. His body was recovered from an embankment just 200 yards from a trackside bothy.

Track workers were not the only ones to die on the line. In February 1924, Mrs M McGill, wife of the stationmaster at Rannoch, was killed when she fell from a moving train in Dalreoch Tunnel, between Helensburgh and Dumbarton. Mrs McGill had been travelling with her young daughter Janet to visit friends in Kirkcaldy when the accident happened. The child alerted railway staff when the train pulled into Glasgow, a search was made, and Mrs McGill's body was recovered. The cause of the accident is not known.

While surfacemen faced physical dangers out on the track, one found his demons closer to home. In October 1935, the LNER received a curious letter from one of its track workers urgently requesting a transfer.

Assigned to rail repairs on Rannoch Moor, Joseph Mullen boarded in one of the two railway cottages at Lubnaclach, which he claimed was haunted.

He wrote: 'Sir. With reference my application of the 15th, I would respectfully point out to you that my grounds for applying for a transfer are that I am unable to stay any longer here as the house that I am in shows signs of it being haunted.

'The extraordinary moving of furniture at night and other signs leaves room for no doubt. I am unable to sleep right with the strain. This has occurred for a time now and I appeal to you to investigate the matter.'

To their credit, the company did investigate, taking statements from the 22-year-old and others living in the house. Mullen claimed that both he and his roommate, a man from the Isle of Skye, had witnessed the nocturnal disturbances.

The two men were initially woken in the night by the sounds of furniture moving and thought at first it had been caused by rats or other animals getting into the room. However, they could find no evidence of this and the noises continued, roughly every other night.

On one occasion, the two men got up, lit the room, and watched as a chair moved across the floor, just four feet in front of their eyes.

Headed by a BR Standard Class 5 and ex-LMSR Class 5 the 7.40am Mallaig to Glasgow train climbs between Tulloch and Corrour, March 1956

'The only person who could have anything to do with it was my mate,' said Mullen. 'But he was as badly scared as I was and he too was glad to get away.'

Mullen dismissed suggestions someone may have been pulling the chair along on a string, saying it was more of a hopping motion.

As the bizarre behaviour continued, Mullen also stated that on occasion he heard footsteps outside the cottage approaching his window.

The LNER, it appears, could find no logical explanation for the reported disturbances but granted Mullen his transfer, to a suburban station in Glasgow.

Whether Mullen and his mate were genuinely plagued by paranormal activity, a poltergeist even, or whether they were just the victims of pranks played by their fellow lodgers will never be known.

Lubnaclach is a lonely spot on a desolate stretch of moor but it does not have a reputation for ghostly goings on. Perhaps, coming from the town, he was just unable to adjust to the eerie silence of a spot where, at night, the slightest sound is amplified. Noises within the room may well have been caused by mice or rats while footsteps outside his window may simply have been the patter of deer hooves. The itinerant chair, however, will remain a mystery.

Over time, to reduce maintenance costs, working practices changed with an increased reliance on motor vehicles to access the track, and fewer men living along the line.

The southern half of the West Highland Railway was relatively easy to reach as it ran parallel with or close to main roads. Between Bridge of Orchy and Tulloch, however, the track was much less accessible.

In the late 1930s, the LNER commissioned Karrier Ltd, of Huddersfield, to build them a solution to this problem. Established in 1908, the company produced light commercial vehicles and buses and had diversified into manufacturing vehicles that could run on both road and rail.

The Karrier E6, number KE 6001, employed by the LNER for track maintenance work on the West Highland Railway

Their road-railer single deck hybrid bus, designed to serve communities that lay just off the railway map, was trialled by the London Midland and Scottish Railway (LMS) in 1931. Cranking flanged wheels down behind each of the four road-going tyres, it took just five minutes to move from road to rail, or vice versa, and the bus could travel at up to 50mph on either.

The weight of the vehicle and its rough ride did not find favour with the LMS directors or the travelling public and the concept never caught on.

However, the LNER saw its potential for their purposes and a goods carrying version with improved chassis, cab and lorry body with fixed sides, detachable tailgate and canvas roof was produced for the company.

It could carry up to 20 men and their tools and could haul up to five wagons of ballast, although in practise it never pulled more than three.

Karrier Ltd was very proud of its association with the LNER and highlighted it in their Road Railer brochure of 1936, stating: 'The vehicle, which travels with equal facility on either road or rail, was the first successful attempt to combine in one vehicle the great safety and cheap operating costs together with the door to door transport advantages of the road vehicle.'

Aquarius Rail Road2Rail 4x4 Land Rover at Rannoch Station (Author)

Deployed in 1938, the vehicle operated between Crianlarich and Mallaig until 1946, aiding inspection of the track and the movement of crews to spots where attention was needed. From Crianlarich it travelled by road to Bridge of Orchy then over the rails of Rannoch Moor to Tulloch. There it continued by road to Fort William and Mallaig. In the same way as locomotives have sand boxes, it was fitted with a sanding device to improve track adhesion.

Karrier Ltd became part of the Rootes Group, manufacturers of the Linwood-built Hillman Imp, but despite road-railer buses being touted as one way to save under-threat branch lines in the 1960s, the technology was never fully adopted.

The road to rail concept was not abandoned by the West Highland Railway. Network Rail use Aquarius Rail Road2Rail 4x4 Land Rovers, designed to transfer from road to rail in less than two minutes, for a variety of tasks on the route, including lineside vegetation inspections, dealing with fallen trees or animal incursions, or repairing rails or trackbed. Based on the Land Rover Defender, the vehicles can carry crew and a one tonne payload and can travel on the tracks at up to 20mph, or 15mph in reverse.

With surfacemen no longer required to live along the line, the railway cottages fell vacant. The house at milepost 52½ was adopted by the Scottish Railway Outdoor Club, established by BR workers, as a base for outings to the Scottish Highlands. Cottages at Rannoch Station were sold

The former gamekeeper's cottage at Lubnaclach being enjoyed by a family on holiday in the 1970s

off while the two cottages at Lubnaclach were demolished.

The neighbouring gamekeeper's cottage remained in use until 1970 when Corrour Estate sold the property with 26 acres of land. During the 1970s and early 1980s the building was shuttered and only occupied periodically as a holiday home by the families and friends of staff at Glasgow University's engineering department.

Danny Phillips recalls staying there in the late 1970s and early 1980s. He said: 'In those days it was possible to get the train guard to stop the train on the line and we could jump out. Or we would throw our bags out the window as we went past so we didn't have to carry them on the walk from Corrour Station to the house.'

He remembers there being three bedrooms upstairs, a bathroom with bath and a kitchen with cooking range. The cottages at Lubnaclach never had mains running water or electricity. Heat was always from open coal or peat fires, light was from paraffin or latterly gas lamps and water was taken from the burn outside.

As part of their stay, visitors were expected to carry out any maintenance that was needed and it was during one such stint of work that the building caught fire. Far from the reach of the fire brigade, the house was destroyed and it now stands as a gaunt ruin.

Ruined gamekeeper's house at Lubnaclach (Author)

9

British Railways

In 1948 Britain's railways, ravaged by the Second World War and in need of significant investment, were nationalised and seven years later a plan to modernise the network, replacing steam with diesel, renewing track and introducing electrification, was deployed.

In the 1960s, Dr Richard Beeching was appointed to streamline the sprawling, loss making, network and his reports *The Reshaping of British Railways*, published in 1963, and *The Development of the Major Railway Trunk Routes*, released in 1965, held little good news for unprofitable lines like the West Highland Railway.

Out of 18,000 miles of railway, Beeching recommended that 6,000 miles or so of mostly rural and industrial lines should be shut down and that some of the remaining lines should be kept open only for freight traffic. A total of 2,363 stations were to go, including 435 already under threat. Both the West Highland Railway and the Far North Line – linking Inverness with Kyle of Lochalsh – were earmarked for complete closure.

Not surprisingly, the move was opposed and, amid concerns over the impact on employment in the West Highlands, the government stepped in with an £8 million loan to fund the construction of a new pulp and paper mill at Corpach. Operated by Wiggins Teape, the plant opened in the Spring of 1966, creating over 700 jobs and prompting a rush of new house building in the area as workers and their families poured in. The company signed a 20-year agreement with British Rail to use the West Highland Railway.

At the time the deal was credited with saving the line but the Corpach mill was never the success the government had hoped. Despite taxpayers backing the venture, the company struggled to get lumber at the price it wanted from the government's own timber agency, the Forestry Commission, and the venture failed in 1991.

Throughout the 1970s and 1980s, the spectre of railway closure lingered with regular 'leaked reports' appearing in the Press indicating the route's days were numbered.

A British Rail Class 27 hauling a Glasgow-bound passenger service stands at Rannoch Station (Sarah Charlesworth)

One threat that targeted the Rannoch Moor portion of the line in particular came when the idea of building a railway between Tulloch and Newtonmore resurfaced. It was first mooted in December 1895 when public meetings were held in both communities to discuss creating a link between Tulloch Station, on the West Highland Railway, and either Newtonmore or Kingussie, on the Highland Railway.

From the outset, the West Highland Railway named its station 'Tulloch for Kingussie' and a daily mail coach service known as the Royal Route plied the A86 by Loch Laggan to Kingussie from the station. Initially horse-drawn, a motorised charabanc took over in May 1915, offering two services a day each way during the summer and one in winter.

After suffering a particularly harsh winter, Laggan residents felt a railway would guarantee the arrival of provisions and mail to the area, aiding farmers and encouraging new industries to the glen.

At the meetings, numerous individuals spoke in support of the proposal and, faced with expensive charges for delivery of coal and other goods on the Highland Railway, the gatherings favoured approaching the West Highland Railway. A committee was formed but ultimately the campaign came to nothing. At the time, the directors of the West Highland Railway and NBR were far more interested in pushing north to Inverness via Fort Augustus and the Great Glen.

A pair of hikers at Corrour, June 1986

In 1974, the Tulloch to Newtonmore idea was resurrected by Lochaber District Council which had noticed fish landings at Mallaig had moved away from the traditional route south over the West Highland Railway to Glasgow and on to Billingsgate and were now going by road to Aberdeen, Fraserburgh and Peterhead, on Scotland's east coast.

One councillor pointed out that rather than build a new railway, fishermen could land their catches at Kyle of Lochalsh and use the branch from there to Inverness to reach the east coast ports.

The proposal did, however, gain some momentum and, in 1978, the Scottish Association for Public Transport (SAPT) published *The Case for a Tulloch to Newtonmore Rail Link*, a document submitted to the Scottish Office and Highlands and Islands Development Board (now Highlands and Islands Enterprise).

Calling for a feasibility study, which was never commissioned, it suggested the route could reduce journey times from Edinburgh to Fort William to just three hours.

The report anticipated that all freight to and from Fort William which, in those days included alumina, aluminium, timber, fish, paper and pulp, could be routed via Newtonmore, taking advantage of an electrified Highland line, something that 40 years on has still yet to happen.

The 20-mile link – estimated to cost around £25 million – would also create a more direct route for alumina trains running from a smelter at Invergordon to the one at Fort William.

Worryingly for the Rannoch Moor stations, the report concluded: 'The Tulloch to Crianlarich section of the West Highland line might be closed or become a pleasure railway.'

At the time, BR was already routing mail and newspapers to Fort William via Newtonmore (a road-going van completed the journey west), the only way it could ensure the day's papers arrived in time for newsagents opening and post arrived in time for the first delivery of the day.

The alumina traffic from Invergordon to Fort William was a healthy revenue stream for the company, even if the route via Glasgow was rather circuitous, but it ended in 1982 when the Invergordon smelter closed, taking some of the wind out of the sails of the proposed Newtonmore to Tulloch link and allowing the railway over Rannoch Moor to sleep easy once again.

A northbound Class 37 hauled passenger train pulls out of Corrour Station, 1987 (Russell Wills)

Threats of closure continued into the 1980s, but the government was always on hand to pump more money into Lochaber. In September 1987, its freight facilities grant scheme awarded over £1 million to British Alcan to help pay for new rail facilities and wagons at the Fort William smelter complex.

The investment was designed to encourage the company to transport the 94,000 tonnes of alumina it imported annually, now from North Blyth, by rail rather than by road. Finished product continued to leave Fort William by road although there was always the hope Alcan would one day transfer this to the track too.

Plans for the Tulloch to Newtonmore link continued to surface from time to time – in January 2016 they appeared in a discussion paper entitled *Scotland 2020-2040 – Major Rail Network Extensions: Potential Roles in an Improved and Inclusive Low Carbon Transport Network* and in 2018 planning engineer James Wegner, originally from Fort William but now living in Australia, revitalised the 1978 SAPT report, once again calling for a feasibility study, a plea which was, once again, ignored.

In the end it was up to operators ScotRail to cut costs to save the line and they did this by dragging the signalling system into the 20th century. But the move would end a decades long tradition and make redundant the men and women who called Rannoch Moor their home. Some, however, were able to adapt, stay on the line and retain their unique way of life.

In 1987 and 1988 the company adopted Radio Electric Token Block (RETB) signalling on the West Highland Railway, replacing the physical tokens handed over by the signalmen or women with an electronic one transmitted from a single control centre in Banavie, near Fort William, direct to the cab of the locomotive via a network of transmitters.

The system was trialled on the Far North line from Inverness to Wick and Thurso in 1978 and on the Dingwall to Kyle of Lochalsh branch, where it was subsequently introduced in 1984. It meant that trains running over a complete route could be controlled from a single point, dispensing with the need to staff signal boxes at intervals along the way.

Gorton was already gone. Even though it was never officially a station, it closed as such in 1964. The signal box, platform and former carriage school were demolished four years later, and the passing loop was lifted.

The last incumbents were Charlie Murray and his stepdaughter Alma Fraser who between them operated a shift system. Charlie opened the signal box at 3.15am in time for The Ghost goods train from Glasgow while Alma took over later in the day, closing the box at 8.15pm if all the trains had come through on time. When the box shut for good, the pair left the employ of the railway.

Gorton's time was, however, not yet up. In 1987, the loop was re-laid as a siding for permanent way trains, controlled by ground levers.

Semaphore signal posts at Corrour and Rannoch were removed in 1985 and the last stationmaster and signaller at Rannoch – Donnie and Eunice McLellan – and at Corrour – Jimmy and Christine Morgan – were made redundant in 1988.

Stationmasters removed, customers bought their tickets on the trains and could use a phone on the platform to contact ScotRail direct if they had any queries.

The move was documented by Scottish Television, which made a half-hour programme entitled *Distant Signals*, highlighting the miserable poverty of the Highlands at the turn of the century which the line was built to address before moving on to focus on the transition from staffed signal boxes and stations to RETB signalling and the introduction of the Class 156 SuperSprinter.

Amid all these changes, Corrour and Rannoch stations were not, as expected, abandoned by their caretakers. At Rannoch, Donnie McLellan had been stationmaster for 35 years, his wife Eunice the equally long-serving signalwoman. Their six children had grown up at Rannoch, taking the train north each day to attend the school at Tulloch.

Donnie recalled that as a young man growing up in Mallaig, job opportunities were limited.

'There was only the railway and the fishing, that was all there was in Mallaig where I was brought up,' he said. 'You either went to join the railway or you went to the fishing, that was the only two jobs you had.

'My wife, she is with me here on the railway as a signal women. My father was on the railway for over 47 years. He was at the building of the Mallaig branch, and my brother was on the railway for 35 years as a loco driver.'

Donnie held his Post Office position until its closure in February 1999 and, reluctant to leave, the couple drew up plans to open a tearoom. ScotRail were keen to find new uses for the vacant station building and, agreeing a peppercorn rent, Eunice, from Glasgow, opened her tearoom, a business that continues to thrive to this day, although now under new owners.

Donnie McLellan in the stationmaster's office at Rannoch Station in January 1982. Note the token equipment and original West Highland Railway clock (Alain Le Garsmeur)

Christine Morgan in the signal box at Corrour, September 1987 (Chris Burton)

At Corrour, stationmaster Jimmy Morgan and his signalwoman wife Christine were also reluctant to go. The couple had been at Corrour since 1978 and had raised their two daughters there. They were keen to maintain a human presence for visitors to the remote outpost. The Post Office had closed in March 1977, but Christine still made Royal Mail deliveries to the youth hostel on her quad bike.

Jimmy had quit a job in electronics in his native Fife to make the move to Corrour, a place he had frequently visited.

He said: 'I knew the place from a long, long time back and I used to come as a youth, fishing and I loved the place,' he said. 'The peace and tranquillity here. If you are a person that can appreciate the nature and the beauty of the surroundings then it is not lonely. Loneliness can only be a frame of mind.'

Christine added: 'It's nice up here. A lot of folk don't like it so peaceful and quiet but we like it. We've got what we want up here and we're quite happy with it. When you come up here you forget the rest of the world exists and you have time to wind down and watch what is going on about you.'

The couple acquired the lease on the signal box, opening a bunkhouse for walkers and climbers – Morgan's Den – and ran a small shop from the kitchen window of their house, selling basic provisions, cigarettes and drink to passing hikers and youth hostel guests.

Corrour Station in the winter of 1982, the prefabricated house occupied by the Morgan's on the right (Steve Duhig)

Jimmy and Christine were keen to maintain a human presence at Corrour, to assist visitors to the station. In such a remote spot, their shop and bunkhouse provided a lifeline for hikers passing through the area and, over the years, they and Corrour Estate staff were involved in many rescues, aiding walkers and climbers who got into difficulties on the moor or surrounding mountains.

While fire had claimed both Corrour Lodge and the gamekeeper's cottage at Lubnaclach, the Morgan's were more fortunate when their home caught light in February 1986. ScotRail swiftly dispatched a special train from Fort William, six firemen equipped with ladders and equipment travelling in a guard's van. When they arrived at the station, Jimmy Morgan had brought the blaze, which started behind the chimney, under control but the firemen ensured it was completely extinguished.

In 1993 ScotRail introduced electricity to Corrour Station, erecting a wind turbine to power platform lighting and electronic departure and arrival display boards. Up until that point the Morgan's and their predecessors had been reliant on Tilley lamps and torches. In later years, bottled gas and a diesel generator housed in a shed adjacent to the platform supplemented supplies before Corrour Estate invested in four new hydro-electric schemes to power its buildings in 2020.

The Morgan's remained at Corrour until the end of 1996 when they retired, returning to Fife.

Staffing costs cut, ScotRail needed to reduce its operating costs and looked to the new Class 156 SuperSprinter diesel multiple unit as a replacement for the 1960s-built Class 37 diesel locomotives which had hauled both passenger and freight trains over the West Highland Railway since the early 1980s when they had taken over from the Class 26 and 27 diesel engines.

First ScotRail Class 156 SuperSprinter 156476, Corrour Summit, 2010 (Author)

Built by Metro-Cammell Ltd in Birmingham, the SuperSprinter was introduced to routes in south-west Scotland in October 1988 and took over on the West Highland Railway on January 23 the following year.

The Class 37s were not completely redundant. They continued to haul summer specials into the mid-1990s, some freight, and the sleeper service to Fort William until 2006.

Hillwalkers disembarking at Corrour, April 2010 (Author)

In 1997 Corrour Estate was purchased by the Rausing family, descendants of Ruben Rausing, one of the founders of the Swedish packaging company Tetra Pak, and in 1999 they set about transforming the station. The house occupied by the Morgan's was demolished and a new structure erected. Initially the building was leased as an independent hostel and café.

In 2010, the previous tenants gone, it was rented to the Scottish Youth Hostels Association and re-opened as a hostel with four bedrooms and a licensed café/restaurant. The venture was short-lived, and it later re-opened as a restaurant/café, its present role.

The disused signal box, the only one of the three remaining on the West Highland Railway (Gorton and Glen Douglas had both been demolished), was classed as a Category C listed building by Historic Scotland in 2013. It decreed that Corrour was an unusual surviving example of a station on the public rail network originally built to serve a private estate.

'Signal boxes are a distinctive and increasingly rare building type that make a significant contribution to Scotland's diverse industrial heritage,' Historic Scotland stated.

'The signal box at Corrour is a non-standard version of the North British Railway's Type 6b, modified to complement the style of the adjacent station waiting room and is therefore also an unusual example of its type.'

Used as both a bunkhouse for walkers and messing point for track maintenance staff, it fell into a poor state of repair in the years following its closure. However, in 2016, thanks to grants totalling £100,000, it was renovated by Network Rail, Corrour Estate and the Railway Heritage Trust and converted into holiday accommodation.

In recent years there has been an interesting reversal of fortunes for the two remaining stations on Rannoch Moor. While Rannoch Station was for many years the busier of the two, attracting the sporting fraternity in droves, Corrour now attracts considerably more passengers than its neighbour.

First ScotRail Class 156 SuperSprinter 156493 at Rannoch Station, January 2010 (Author)

In the 1930s, Rannoch was the busiest intermediate station on the line, attracting over 11,000 passengers a year. It now draws, on average, between 7000 and 8000 people a year. Corrour, by contrast, consistently ranks as the fourth most popular halt on the West Highland Railway, behind Arrochar & Tarbet, Crianlarich and Helensburgh Upper, and is the busiest intermediate station on the northern half of the route, not bad for a remote outpost that did not figure on the original plans for the line and only then started life as a private halt.

On track between Corrour and Tulloch, English Welsh & Scottish Railway (EWS) Class 67 diesel-electric loco 67009 hauling the Caledonian Sleeper north to Fort William, June 2012 (Author)

In 2019/20, 12,630 people bought a ticket to travel to or from Corrour, a slight dip on previous years. Most of these visitors are hillwalkers. A trio of Munros – Scottish mountains over 3000 feet high – encircle Loch Ossian.

Carn Dearg, Sgor Gaibhre and Ben na Lap ensure a steady flow of Munro-baggers while many simply enjoy a stroll around the loch itself, a very pleasant outing. Leum Uilleum, west of the station, is well worth climbing while well-walked rights of way lead west to Fort William, east to Dalwhinnie via Ben Alder and south to Loch Rannoch.

In an effort to cater for those drawn to the line and its stations by outdoor recreation, ScotRail introduced 'Active Travel' coaches to the route in 2021. These converted Class 153 single-car diesel multiple units feature racks for bicycles, more luggage space for larger, bulkier sports equipment such as skis, power sockets and wi-fi, and operate coupled to existing Class 156s.

Popular culture has also played its part in drawing visitors to the station, most notably a scene from the 1986 Danny Boyle directed film *Trainspotting* where the main characters travel to Corrour, find there is nothing there and, after discounting an ascent of Leum Uilleum, catch the next train home.

Corrour Station, June 2008, showing the wooden shelter. The gravel has since been replaced with asphalt to improve access for passengers with disabilities (Author)

A more obscure film, in which Corrour Station plays a prominent role, is the 1986 Dutch movie *De Wisselwachter* (*The Pointsman*). Opening on the Forth Railway Bridge, some atmospheric carriage window shots of a misty Moor of Rannoch lead to Corrour where a French woman accidentally alights.

Forced to wait for the next train, which never comes, she develops an unlikely relationship with the pointsman, despite the fact neither speaks the other's native tongue. There are plenty of shots of the station, the moor and a grey Class 37 diesel engine that trundles by from time to time.

On a more mainstream note, the *Harry Potter* series, which made good use of the West Highland Railway during filming, did a brief bit of shooting on the line south of Corrour Station.

Today, both Corrour and Rannoch are passenger only stations. Services generally run three times a day in each direction, but less frequently on Sundays (twice each way in summer, but just once in winter). They are operated by ScotRail, the franchise currently held by Dutch public transport company Abellio.

This ends in March 2022 when ScotRail will be nationalised, services thereafter being operated within the public sector by a company owned and managed by the Scottish Government.

The stations are also served by the Caledonian Sleeper, one of only two sleeper services in Britain, which runs daily (except Saturday) from London Euston to Fort William, Aberdeen and Inverness.

The sleeper train was for many years operated by ScotRail but, from April 2015, the franchise has been held by Serco, an outsourcing company that specialises in public service work. In 2019, Serco invested in a new fleet of Spanish-built Mark 5 carriages to replace the existing rolling stock, offering an improved level of comfort.

Departing London Euston, the 16-carriage train stops at Watford Junction, Crewe and Preston on its journey up the West Coast Main Line to Edinburgh Waverley. There it splits into three parts, the front portion proceeding to Fort William with stops at Glasgow Queen Street, Dalmuir, Dumbarton Central, Helensburgh Upper, Garelochead, Arrochar &

Tarbet, Ardlui, Crianlarich, Upper Tyndrum, Bridge of Orchy, Rannoch, Corrour, Tulloch, Roy Bridge and Spean Bridge.

At the time of writing, the only freight service on the West Highland Railway is the GB Railfreight-hauled alumina train which runs six days a week to the smelter in Fort William, now owned and operated by ALVANCE British Aluminium.

Acknowledgements

Photographs

Every effort has been made to trace and contact the copyright holders of images reproduced within this book and acknowledge them as such. In some instances this has not been possible. If you hold the copyright to any of the uncredited images within this book, please contact the author (jimcarron@gmail.com).

Images on pages 20, 44, 95, 97 and 102 are licensed for re-use under the Creative Commons Attribution-ShareAlike 2.0 Licence

Maps

Maps on pages 22 (Caledonian Railway Company), 34, 35, 37, 40, 42 and 48 (Ordnance Survey) are reproduced with the permission of the National Library of Scotland. The map on page 4 incorporates mapping © OpenStreetMap contributors. Contains Ordnance Survey data © Crown copyright and database right 2010-19. Data is available under the Open Database License. For more information, visit openstreetmap.org.

Archives

The British Newspaper Archive

National Library of Scotland

National Records of Scotland

Scotland's People

Bibliography

The Story of the West Highland, George Dow (LNER, 1944)

The West Highland Railway, John Thomas (David St John Thomas, 1984)

Periodicals

Aberdeen Free Press; Aberdeen Press & Journal; Blackwell's Magazine, Edinburgh; Daily Telegraph & Courier; Dundee Courier; Dundee Evening Telegraph; Glasgow Herald; London Evening Standard; London Illustrated News/The Sphere; Railway Magazine; Port Glasgow Express; and The Scotsman, Edinburgh.

About the Author

James Carron is a freelance writer based in Scotland.

Contact jimcarron@gmail.com

Other Books by James Carron

Secret Scotland

amenta.ink

Fifty unusual and offbeat attractions, quirky curiosities and hidden gems, secluded and less well-known spots that await discovery. Most can be visited at any time of the year, day or night, with no booking required or admission charged.

Highland Hermit – The Remarkable Life of James McRory Smith

amenta.ink

Biography of James McRory Smith who lived for over 30 years at Strathchailleach, one of the most remote cottages in the Britain Isles. An inspiring account of a modern day hermit, the book offers a rare insight into an alternative way of life, one far removed from the norm. Both social history and true-life story of adventure and survival.

For more information on Amenta Publishing titles, available in both ebook and paperback formats from Amazon, visit www.amenta.ink

Printed in Great Britain
by Amazon